teach yourself®

astronomy

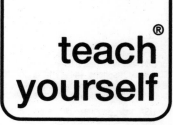

astronomy
sir patrick moore

Launched in 1938, the **teach yourself** series grew rapidly in response to the world's wartime needs. Loved and trusted by over 50 million readers, the series has continued to respond to society's changing interests and passions and now, 70 years on, includes over 500 titles, from Arabic and Beekeeping to Yoga and Zulu. What would you like to learn?

be where you want to be with **teach yourself**

For UK order enquiries: please contact Bookpoint Ltd, 130 Milton Park, Abingdon, Oxon, OX14 4SB. Telephone: +44 (0) 1235 827720. Fax: +44 (0) 1235 400454. Lines are open 09.00–17.00, Monday to Saturday, with a 24-hour message answering service. Details about our titles and how to order are available at www.teachyourself.co.uk

For USA order enquiries: please contact McGraw-Hill Customer Services, PO Box 545, Blacklick, OH 43004-0545, USA. Telephone: 1-800-722-4726. Fax: 1-614-755-5645.

For Canada order enquiries: please contact McGraw-Hill Ryerson Ltd, 300 Water St, Whitby, Ontario, L1N 9B6, Canada. Telephone: 905 430 5000. Fax: 905 430 5020.

Long renowned as the authoritative source for self-guided learning – with more than 50 million copies sold worldwide – the **teach yourself** series includes over 500 titles in the fields of languages, crafts, hobbies, business, computing and education.

British Library Cataloguing in Publication Data: a catalogue record for this title is available from the British Library.

Library of Congress Catalog Card Number: 94-68413

First published in UK 1995 by Hodder Education, part of Hachette Live UK, 338 Euston Road, London, NW1 3BH.

First published in US 1995 by The McGraw-Hill Companies, Inc.

This edition published 2008.

The **teach yourself** name is a registered trade mark of Hodder Headline.

Typeset by Transet Limited, Coventry, England.
Printed in Great Britain for Hodder Education, an Hachette Livre UK Company, 338 Euston Road, London NW1 3BH, by Cox & Wyman Ltd, Reading, Berkshire.

The publisher has used its best endeavours to ensure that the URLs for external websites referred to in this book are correct and active at the time of going to press. However, the publisher and the author have no responsibility for the websites and can make no guarantee that a site will remain live or that the content will remain relevant, decent or appropriate.

Hachette Livre UK's policy is to use papers that are natural, renewable and recyclable products and made from wood grown in sustainable forests. The logging and manufacturing processes are expected to conform to the environmental regulations of the country of origin.

Impression number   10 9 8 7 6 5 4 3 2 1
Year   2012 2011 2010 2009 2008

contents

foreword

Astronomy today is a fast-moving science, and much has happened since the publication of the last edition of this book. I have therefore done my best to bring the text up to date. I hope that the result is acceptable!

Sir Patrick Moore, Selsey, January 2008.

about the author

Sir Patrick Moore has been interested in astronomy since the age of six, and has specialized in studies of the Moon. He has his private observatory at Selsey in Sussex, and has presented the *Sky at Night* programme on BBC Television since 1957.

acknowledgements

The Publishers would like to thank the following for kindly allowing the use of their photographs in the plate section of this book: Plate 1, Chris Doherty; Plate 2, Bruce Kingsley; Plate 3, Jeremy Rundle; Plate 4, Pete Lawrence; Plate 5, Bruce Kingsley; Plate 6, NASA; Plate 7, Russian Academy of Sciences; Plate 8, Bruce Kingsley; Plate 9, Pete Lawrence; Plate 10, Ian Sharp; Plate 11, Ian Sharp; Plate 12, Pete Lawrence.

preface

Astronomy is both the easiest and the most difficult of all the sciences. It is the easiest because anyone can take a really intelligent interest in it, and can carry out useful observations without the need for technical training or expensive equipment. It is the most difficult because it involves concepts which nobody can fully understand.

It is also a fast-moving science; new discoveries are made with bewildering rapidity, and new theories are advanced to try to explain them. Over the past decade many seemingly well-established theories have been swept away; the astronomy of 2008 is very different from that of 1998.

What I aim to do here is to present a general survey for the newcomer *with no previous knowledge*. Moreover, the mathematics I have used involves nothing more taxing than simple addition, subtraction, multiplication and division. I do not set out to present a comprehensive guide to observing – the books listed in the 'Taking it further' section will do that – but if you follow what I have to say, I hope that you will feel inclined to take matters further.

Astronomy is still just about the only science in which amateurs can make useful contributions. Their work is genuinely appreciated by professional astronomers, and it is fair to say that dedicated amateurs can carry out work that the professional has no time to do, no wish to do, or is unable to do. To give just one example: amateurs have a fine record in discovering new comets and exploding stars. It is true that some amateurs now use complicated electronic equipment, but others are content with very modest telescopes. Even with the naked eye alone, there is plenty that can be achieved. Astronomy also has the advantage that it can take up as much, or as little, of your free time as you wish.

To become an amateur astronomer you do not need any concrete qualifications. I would suggest joining one of the many astronomical societies – either national or local – as you will not only learn more about astronomy, but will make a host of new friends. Astronomy is surely the best of all hobbies.

So let us begin at the beginning.

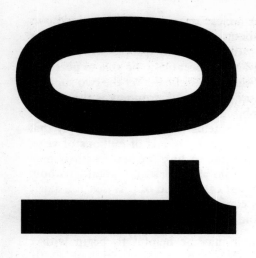

01

introducing astronomy

In this chapter you will learn:
- the basic facts about astronomy
- the definitions of the most important terms.

I would like to start with a clear statement: 'Everybody knows what astronomy is all about.' Unfortunately, this statement is not true. Even today I still find many people who confuse astronomy with astrology, and expect me to be able to peer into the future, probably by using a crystal ball. Yet the two subjects could not possibly be more different. Astronomy is the science of the sky and all the objects in it, from the Sun and Moon through to the remote star-systems which are so far away that we see them as they used to be thousands of millions of years ago. Astrology, which attempts to link the stars with human character and destiny, is a relic of the past; it is totally without foundation, and the best that can be said of it is that it is fairly harmless so long as it is confined to seaside piers, circus tents and the columns of tabloid newspapers.

Having made this clear, it may be helpful to begin with a celestial roll-call. Obviously we must start with our Earth, which was once thought to be the most important body in the entire universe, but which we now know to be an insignificant planet moving around an insignificant star – the Sun. The distance between the Sun and the Earth is 149.5 million km (93 million miles), and it takes one year for the Earth to complete a full circuit: more precisely, 365.2 days. The Earth spins around once in approximately 24 hours, and is surrounded by a layer of atmosphere which extends upwards for several hundreds of miles even though most of it is concentrated at low levels.

At 149.5 million km (93 million miles) from us the Sun seems a long way away, but in astronomy we have to deal with immense distances and vast spans of time. Nobody can really understand figures of this sort – certainly I have no proper appreciation of even 1 million km (0.6 million miles) – but we know that the values are correct, and we simply have to accept them.

Just as the Earth is an ordinary planet, so the Sun is an ordinary star. All the stars you can see on any clear night are themselves suns, some of them far larger, hotter and more luminous than ours. They appear so much smaller and dimmer only because they are so much further away. Represent the Earth–Sun distance by 2.5 cm (1 inch), and the nearest star will be over 6.5 km (4 miles) away. The Pole Star, which many people recognize (and about which I will have more to say in the next chapter) will have to be taken to a distance of 11,260 km (7000 miles). It is at least 6000 times more luminous than the Sun, and yet it seems by no means the most brilliant star in the night sky.

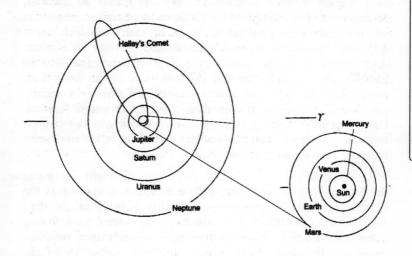

Figure 1.1 Plan of the Solar System, including Halley's Comet

Around the Sun move eight planets, of which the Earth comes third in order of distance. Mercury and Venus are closer to the Sun than we are; Mars, Jupiter, Saturn, Uranus and Neptune are further away. Unlike the stars, planets have no light of their own, but shine only because they reflect the rays of the Sun. If some malevolent demon suddenly snatched the Sun out of the sky, the planets (and the Moon) would vanish from sight, though naturally the other stars would be unaffected.

The planets move around the Sun at different distances and in different periods. Even a casual glance at a plan of the Sun's family, or Solar System, shows that it is divided into two well-marked parts (see Figure 1.1). First we have the four small planets (Mercury to Mars) and then a wide gap, followed by the four giants (Jupiter to Neptune). The wide gap between the paths or orbits of Mars and Jupiter is occupied by a swarm of very small worlds known variously as minor planets, planetoids and (more commonly) asteroids.

The revolution periods of the planets range from 88 days for Mercury to almost 165 years for Neptune. They are very different sizes; Jupiter is over 143,000 km (89 000 miles) in diameter, Mercury only just over 4800 km (3000 miles). Because the planets are relatively near neighbours, some of them can look most imposing. Venus, Jupiter and Mars at their best are far brighter than any of the stars, while Saturn is prominent enough to be conspicuous, and Mercury can often be seen in the twilight or dawn sky. All these planets have been known since very ancient times, while the others were discovered much more recently: Uranus in 1781 and Neptune in 1846. Uranus can just be seen with the naked eye if you know where to look for it, but to observe Neptune you need optical aid.

What, then, of the Moon, which dominates the night sky just as the Sun is king of day? Officially the Moon is known as the Earth's satellite; it moves around us at a distance of less than 402 250 km (250 000 miles), and it is little more than 3200 km (2000 miles) across. Like the planets, it depends upon reflected sunlight. Obviously the Sun can illuminate only half of the Moon at any one time, and this is why we see the regular phases, or apparent changes of shape, from new to full; everything depends upon how much of the Moon's sunlit hemisphere is turned in our direction. It is an airless, waterless, lifeless world, but it is our faithful companion in space, and stays together with us in our never-ending journey around the Sun. It takes just over 27 days to complete one orbit of the Earth, though for reasons to be explained later the interval between one new moon and the next is 29½ days.

Other planets have satellites of their own – over 60 each for Jupiter and Saturn – but in some ways the Earth–Moon system is unique, and it may be better to regard it as a double planet rather than as a planet and a satellite.

Among other members of the Sun's family are comets, which have been referred to as 'dirty snowballs'. They are flimsy, wraithlike things; with even a major comet the only substantial part is the nucleus, made up of a mixture of ice and 'rubble' and seldom more than a few kilometres across. Like the planets, comets travel around the Sun, but whereas the orbits of the planets are almost circular those of the comets are, in most cases, very eccentric. When a comet nears the Sun, its ices begin to evaporate, and the comet may produce a gaseous head and a long tail; when the comet retreats once more into the cold depths of the Solar System, the head and tail disappear, leaving only the inert nucleus. The only bright comet to appear regularly is Halley's (named in honour of the second Astronomer Royal, Edmond Halley), which comes back every 76 years, and last paid a visit to the Sun in 1986. Really brilliant comets have much longer periods, so that we cannot predict them. Now and then a comet meets with a dramatic end, as in July 1994 when a comet known as Shoemaker–Levy 9 crashed to destruction upon the planet Jupiter.

Note that a comet moves well above the top of the Earth's air, and is a long way away, so that it does not seem to crawl quickly across the sky; if you see a shining object which is shifting perceptibly, it cannot be a comet.

As a comet travels, it leaves a trail of 'dust' behind it. If one of these dusty particles dashes into the upper atmosphere, it will have to push its way through the air-particles, so that it becomes heated by friction and burns away in the streak of luminosity which we call a meteor or shooting-star. Therefore, a shooting-star has absolutely no connection with a real star; it is a tiny piece of débris, usually much smaller than a pin's head, which we see only during the last few seconds of its life before it burns away completely, ending its journey to the ground in the form of ultra-fine dust.

Larger bodies, not associated with comets, may survive the full drop without being burned away, and are then known as meteorites. Some of them may make craters; for example, the Arizona Crater in the United States was certainly formed by an impact more than 50 000 years ago. Meteorites come from the asteroid belt, and in fact there is no difference between a large meteorite and a small asteroid; it is all a question of terminology. In recent years several tiny asteroids, much less than 1.6 km (1 mile) across, have been known to pass by us at less than half the distance of the Moon. One of these cosmical midgets is believed to have been no more than 9 m (30 ft) in diameter.

The planets move around the Sun in very much the same plane, so that if we draw a plan of the Solar System on a flat piece of paper it is not far wrong. This does not apply to the comets or asteroids, and indeed some comets move around the Sun in a 'wrong-way' or retrograde direction; Halley's Comet is one of these. In addition, there is a great deal of material spread thinly along the main plane of the system, which shows up when lit by the Sun and produces the lovely cone-shaped glow that we call the Zodiacal Light.

The planets were first identified because, unlike the stars, they shift in position from one night to another; indeed, the word 'planet' really means 'wanderer'. The stars are so remote that the individual or proper motions are very slight, and the star-patterns or constellations seem to remain to all intents and purposes unchanged over periods of many lifetimes. Board Dr Who's time machine and project yourself back to the age of William the Conqueror, or Julius Caesar, or even Homer: the constellation patterns will appear practically the same as they do now. It is only the members of the Solar System which move more obviously against the background, and even then the Sun, Moon and principal planets keep strictly to a band around the sky which we call the Zodiac.

The constellations which we use today have come down to us from the Greeks. The last great astronomer of Classical times, Ptolemy, gave a list of 48 constellations, all of which are still to be found on our maps even though their boundaries have been modified and new groups added. The Greek names (suitably Latinized) commemorate the mythological gods and heroes, together with living creatures and a few inanimate objects. Thus we have Orion (the Hunter), Hercules (the legendary hero), Ursa Major (the Great Bear), Taurus (the Bull) and others. Probably the best-known constellations visible from northern countries are the Great Bear and Orion, both of which are distinctive; the Bear never sets over Britain, though Orion is out of view during the summer because the Sun is too close to it in the sky. Yet it is important to remember that the constellation names and patterns mean nothing at all, because the individual stars are at very different distances from us.

Star distances are so great that ordinary units of measurement, such as the kilometre and the mile, are inconveniently short (just as it would be cumbersome to give the distances between London and New York in centimetres). Luckily, nature has provided us with an alternative. Light does not move instantaneously; it flashes along at 299 000 km (186 000 miles) per second, so that in a year it can cover approximately

9 500 000 million (6 million million miles). It is this distance which is termed the light-year. The nearest star to the Sun is over four light-years away; the Pole Star, 680 light-years; Rigel, the brilliant white star in Orion, about 900 light-years. It follows that once we pass beyond the Solar System, our view of the universe is bound to be out of date. We see Rigel as it used to be 900 years ago, and if it were suddenly extinguished we would not know for another 900 years. Some recent measures give different distances for many of the brilliant stars. In this book I have followed the values given in the authoritative Cambridge catalogue. On the other hand, light takes only 8.6 minutes to reach us from the Sun, and can leap from the Moon to the Earth in only 1.25 seconds.

The essential fact is that the stars in any particular constellation are not genuinely associated with each other; they merely happen to lie in much the same direction as seen from Earth, so that we are dealing with nothing more significant than a line-of-sight effect. In Orion, for example, the two leading stars are the orange-red Betelgeux and the white Rigel. Betelgeux is 310 light-years away from us, Rigel 900, so that Rigel is much further away from Betelgeux than we are – and if we were observing from a different vantage point, Betelgeux and Rigel might well lie on opposite sides of the sky. In fact, a constellation name means absolutely nothing. Among early civilizations, the Chinese and the Egyptians had constellations of their own. If we had followed, say, the Egyptian system we would have had a Cat and a Hippopotamus instead of a Bear and a Bull, though, needless to say, the stars would have been exactly the same.

Naturally, this line-of-sight principle applies also to the wandering planets. When we say that, for example, Mars is 'in Taurus', what we mean is that the planet is seen against a background of totally unconnected stars which we happen to call the Bull, even though the stars in this particular group do not form any distinct pattern, and certainly do not conjure up the impression of a bull or any other creature.

Although the constellations do not change obviously over many centuries, they are not absolutely permanent. The stars are not fixed in space; they are rushing about in all sorts of directions at all sorts of speeds, and it is only their remoteness which makes them appear sluggish (just as a jet moving against the clouds will seem to move much more slowly than a sparrow flying around at tree-top level). Eventually the constellation patterns will change, and if we could come back in, say, 50 000 years, the night sky would seem unfamiliar.

Our Sun is a member of a system which we call the Galaxy. It contains about 100 000 million stars altogether, arranged in a form which is usually likened to a double-convex lens but which I prefer to compare with the shape of two fried eggs clapped together back to back. The Sun, together with the Earth and other members of its family, lies not far from the main plane of the Galaxy, but is nowhere near the centre of the system, which is almost 30 000 light-years from us; the overall diameter of the Galaxy is of the order of 100 000 light-years. When we look along the main plane we see many stars in almost the same direction, and this is what produces the Milky Way, that lovely, glowing band which stretches across the night sky from one horizon to the other. The Milky Way is made up of stars, seemingly so close together that they are in danger of colliding with each other, but once again appearances are deceptive. The stars are widely spaced, and in average regions of the Galaxy, at least, direct collisions must be very rare indeed. Even 'close encounters' occur seldomly.

Here and there we find whole clusters of stars, some of them loose and irregular in form, others symmetrical and globular. Of the open clusters the most famous is that of the Pleiades or Seven Sisters; a familiar sight in the night sky for two months either side of Christmas. Globular clusters are remarkably regular, and may contain up to a million stars.

Also in the Galaxy we find clouds of gas and dust known as nebulae. A few of them are visible with the naked eye, and telescopes show vast numbers. Nebulae are stellar birthplaces, where fresh stars are being formed from the tenuous interstellar material. Around 5000 million years ago, our Sun was born inside a nebula in just this way.

Our Galaxy is by no means the only one. Modern equipment can reveal at least 1000 million others, but all are a long way away. Only a few are within a million light-years of us, and most are at immense distances. In the northern sky, the most famous outer galaxy is the Great Spiral in the constellation of Andromeda, which is dimly visible with the naked eye but very easy to see with binoculars. It is 2.2 million light-years away, and is larger than our Galaxy, with more than our quota of 100 000 million stars. Telescopes show that it is spiral in form, like a cosmic Catherine-wheel, and this is no surprise; there are many spiral galaxies, including our own. If we could see our Galaxy from 'above' or 'below', the spiral shape would be very evident. The Sun lies close to the edge of one of the spiral arms.

Even the Andromeda system is a relatively near neighbour, and most galaxies are so remote that they appear comparatively faint. Today we know of galaxies which are well over 10 000 million light-years away, so that we see them as they used to be when the universe was young. We also know that all the galaxies except those of our own particular 'local' group are racing away from us – and the further away they are, the faster they are going. This does not mean that we are particularly unpopular, or that we are in any privileged position. The entire universe is expanding, and every group of galaxies is receding from every other group.

How big is the universe, and how was it born? These are the most fundamental of all questions. I will discuss them later in this book, but I cannot hope to provide satisfactory answers. All we can really say is that the universe in its present form is 13 700 million years old. I avoid using 'billion' because the US billion – 1000 million – differs from the old English value of 1 million million. The US billion is now almost universally accepted, but there is still a possibility of confusion. We can set a limit to the size of the observable universe, but this is by no means the whole story.

I hope that this 'roll-call' is of help, and at the end of the book I have added a glossary which may be useful. There is one other point which is worth making here. On countless occasions during the past 40 years or so people have said to me: 'What has happened to the stars? They used to be so much brighter than they are now; have they faded?' The answer, of course, is that they have not – but our skies have become brighter. Glaring artificial illuminations cast a sheen over the entire heavens, and light pollution has become a very serious problem, not only to astronomers. Sadly, anyone who lives inside a city or a densely-populated area will have at best a very poor view, and will certainly never be able to see the beauty of the Milky Way.

Efforts are being made to tackle the problem. It is not true to say, as some people have done, that astronomers want to 'put out the lights'; in the modern world, with the general breakdown in law and order, well-lit streets are essential. The aim is to provide lights which shine down, not up, and encouraging progress is being made; but there is a long way to go, and it would indeed be tragic if future generations were deprived of the glory of the night sky.

02

the spinning sky

In this chapter you will learn:

- about the 'celestial sphere'
- how the positions of celestial bodies are given in the sky
- the celestial equivalents of latitude and longitude on the Earth.

It is easy to imagine that we are living on a steady, motionless Earth, but this is certainly not true. The Earth is turning on its axis once in 24 hours, so that we are being whirled about; the Earth is moving around the Sun at an average speed of about 106 000 kmh (66 000 mph), and the Sun itself is travelling around the centre of the Galaxy at over 209 km (130 miles) per second, taking us with it.

Our path around the Sun is not perfectly circular; like the orbits of all the other planets it is slightly elliptical, and our distance ranges between 147 million km (91.4 million miles) in December out to 152 million km (94.6 million miles) in June. The seasons have very little to do with the changing distance. The Earth's axis of rotation is tilted to the perpendicular to the orbit by 23½ degrees, and in December the north pole is tipped away from the Sun, so that it is winter in the northern hemisphere; in June it is the south pole which is tipped away, and the northern part of the world enjoys its summer (see Figure 2.1).

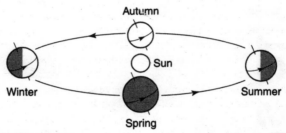

Figure 2.1 The seasons

Ancient peoples believed the Earth to be surrounded by a solid invisible crystal sphere, whose centre was coincident with the centre of the Earth's globe. In some ways it is still very convenient to visualize this 'celestial sphere', and to assume that it really does revolve around the Earth in a period of 24 hours, carrying the Sun, Moon, planets and stars with it. We can then fix the positions of the celestial objects on the sphere, much the way in which we fix positions on the Earth by means of latitude and longitude.

The first step is to define the poles of the sky. These lie in the direction of the Earth's axis of rotation, as shown in Figure 2.2. The north celestial pole is marked within 1 degree by the brightish star Polaris, in the constellation of Ursa Minor, the Little Bear; the opposite pole is not so favoured, and the south pole star, Sigma Octantis, is none too easy to see with the naked eye.

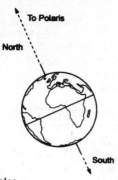

Figure 2.2 The celestial poles

It is easy to see that as the Earth spins, the Pole Star seems to remain motionless in the sky (or virtually so), with everything else turning around it. This is not to say that Polaris is important in itself; it owes its eminence to the fact that it chances to lie in the direction of the Earth's axis. This has not always been so, because of an effect known as precession (see Figure 2.3). The Earth is not a perfect sphere; its diameter is 12 753 km (7926 miles) as measured through the equator, but only 12 711 km (7900 miles) as measured through the poles. The Sun and Moon pull on this equatorial bulge, and the result is that the Earth's axis wobbles slightly in the manner of a gyroscope which is running down. A gyroscope will wobble or 'precess' in a few seconds; the Earth's axis takes 25 800 years to complete a full turn, but it means that the position of the celestial pole changes slowly but steadily. At the time when the Egyptian Pyramids were being built, the pole lay near a much fainter star, Thuban in Draco (the Dragon); it is now near Polaris, but it will not stay there, and in 12 000 years from now the north pole star will be Vega in Lyra (the Lyre), one of the most brilliant stars in the sky.

The altitude of the celestial pole above the horizon depends on the latitude of the observer on the Earth's surface. For example, if you find that Polaris is 60 degrees in altitude, you will know that your latitude is 60 degrees north, about the same as that of St Petersburg. (For the moment, I propose to simplify matters by assuming that Polaris lies exactly at the polar point.) On the equator, Polaris will lie on the horizon – altitude 0 degrees – and from southern latitudes it can never be seen at all; we have to make do with the obscure Sigma Octantis.

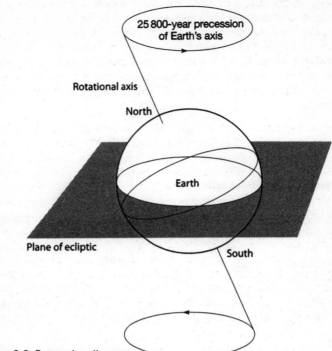

Figure 2.3 Precession diagram

If a star is sufficiently close to the pole, it will be circumpolar; that is to say, it will remain above the horizon all the time. For example, consider Ursa Major, the Great Bear. It is in the far

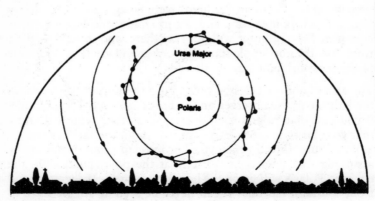

Figure 2.4 Circumpolar star diagram

north of the sky, and from Britain it never sets; it merely goes around and around the pole, as shown in Figure 2.4, whereas Regulus in Leo (the Lion), further away from the pole, spends part of its 24-hour circuit below the horizon. Go to a more southerly latitude, and you will find that the Bear is no longer circumpolar, though to compensate for this it becomes possible to see stars which are too far south in the sky to rise over Britain.

Just as the Earth's equator cuts the world in two, so the celestial equator cuts the sky in two; it is simply the projection of the Earth's equator on to the celestial sphere. On Earth, latitude is defined as the observer's angular distance north or south of the equator, as measured from the centre of the globe; approximately 51 degrees north for London, 38 degrees north for Athens, 12 degrees south for Darwin in Australia, 34 degrees south for Cape Town, and so on. Obviously, the latitude of the north pole is 90 degrees north, and that of the south pole 90 degrees south. To fix your position you need to know your longitude as well as your latitude, and here we measure the angle from the great circle on the Earth's globe which passes through both poles and also Greenwich Observatory in Outer London, as shown in Figure 2.5. Greenwich was accepted as the

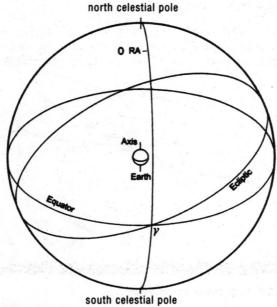

Figure 2.5 Celestial sphere

zero for longitude during the 1880s, when international agreement was much easier to obtain than it is now.

Now let us go back to the celestial sphere, and define the equivalents of latitude and longitude. Declination is simply the angular distance of the star (or other body) north or south of the celestial equator – for example 39 degrees north or +39° for Vega and 53 degrees south, or –53°, for Canopus in the constellation of Carina, the Keel. (Again I am using round numbers; the actual value for Vega is +38°47'01".) This is the celestial equivalent of latitude, and presents no problems at all, but the equivalent of terrestrial longitude is rather less straightforward.

The apparent yearly path of the Sun around the sky is called the ecliptic; it may be defined as the projection of the Earth's orbit on to the celestial sphere. Since the axis is tilted to the perpendicular by 23½ degrees (more precisely 23°.44), this is also the angle between the ecliptic and the celestial equator. Therefore, the Sun's declination can range between 23½ degrees north and 23½ degrees south; the points of greatest declination are known as the solstices (see Figure 2.6). The

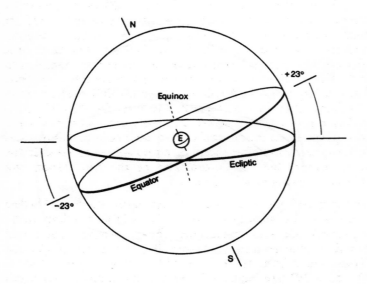

Figure 2.6 Ecliptic and equator

actual dates vary slightly because of the vagaries of our calendar, but not by more than a day or so; in 2008 the summer solstice fell on 20 June and the winter solstice on 21 December.

During its annual journey against the stars the Sun must cross the equator twice, once when travelling from south to north, around 20 March, and once when travelling from north to south, around 23 September. These two points are known as the equinoxes. They are not marked by any bright stars, and of course precession means that they shift slowly and inexorably, because a change in the position of the celestial pole also means a change in the position of the celestial equator. In Pyramid times the March or vernal equinox lay in the constellation of Aries, the Ram, and it is still known as the First Point of Aries, even though precession has long since moved it into the adjacent constellation of Pisces, the Fishes. It is the First Point of Aries which we take as the zero for the celestial equivalent of longitude, known as right ascension.

Right ascension (R.A.) is defined as the angular distance between the object and the First Point of Aries, but, rather confusingly, it is always given in units of time (see Figure 2.7). As the Earth spins, a celestial body will rise in an easterly direction and set towards the west (unless it is circumpolar, when it will remain in view all the time). When it reaches its highest point above the horizon, it is said to culminate. The First Point of Aries must culminate once in every 24 hours, though naturally this often happens during daylight. The R.A. of a celestial body is given by the time-lapse between the culmination of the First Point of Aries and the culmination of the body. Thus Altair, a bright star in Aquila (the Eagle), culminates 19 hours 51 minutes after the First Point has done so; therefore, the R.A. of Altair is 19h 51m. Because the stars remain in virtually the same positions on the celestial sphere, their right ascensions and declinations do not alter except by a very slight annual shift due to precession, but the co-ordinates of the Sun, Moon and planets are changing all the time.

Given the declination of a star, you can easily work out whether it can be seen from your observing site. All that you have to do is to work out your co-latitude, which is simply your actual latitude subtracted from 90 degrees. My own home, at Selsey in Sussex, lies at latitude 51 degrees N, so that my co-latitude is 90 − 51 = 39° (see Figure 2.8). This means that any star north of declination +39° will never set, and any star south of declination −39° will never rise. I can always see Deneb in Cygnus, the Swan (declination +45°), but I have no sight of Canopus (declination −53°). Invercargill, in New Zealand, lies at latitude 46 degrees south, so that from here Canopus is circumpolar and Deneb can never be seen.

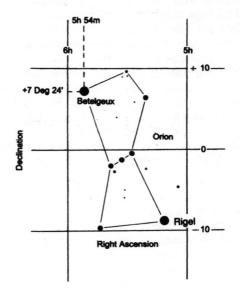

Figure 2.7 Right ascension (R.A.)

When the Sun is north of the celestial equator, it is circumpolar from the north pole of the Earth, and this is why there is a six-months' 'day' there. When the Sun moves south of the equator it is the turn of the south pole to have its 'midnight sun', and at the moment a major observatory is being set up right on the south pole itself. The site is particularly favourable astronomically; it will disturb nothing – even penguins avoid Central Antarctica! – and there are high hopes for the new observatory, though it must be admitted that conditions there will be somewhat chilly.

Certainly the celestial sphere is a useful concept, and at least it allows us to clarify the way in which we fix the positions of the objects in the sky – even though we no longer believe, as our forebears did, that the world is surrounded by a solid sphere.

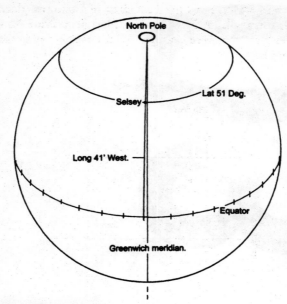

Figure 2.8 Greenwich meridian

03

sky-watchers

In this chapter you will learn:
- about the history of astronomy from the old sky-watchers to the huge telescopes and space-probes of today.

A famous American was wrong when he once said that 'History is bunk'. It is not; and I doubt whether anyone can really appreciate a subject without knowing at least a little about its background. The story of astronomy goes back to our cave-dwelling ancestors, and to summarize it in a single chapter is a truly Himalayan task, but at least I can say a little about some of the main highlights.

Originally it was assumed that the Earth must be both flat and motionless. This was natural enough – after all, the Earth does look flat apart from local irregularities such as hills and valleys, and we have no sensation of being whirled along at breakneck speed. It was also assumed that the Sun, Moon and stars had been created for the specific benefit of humankind, and, as we have noted, the sky was thought to be solid. (According to the Egyptians it was in fact the body of a goddess with the rather appropriate name of Nut, who was crouched over the flat earth in a decidedly inelegant position.) The sky revolved around the Earth once in 24 hours, carrying all the celestial bodies with it.

It is only with the development of writing that we have any positive information to guide us, but it seems that the constellation patterns date back to 3000 BC or even earlier. We cannot be sure, and neither do we know which peoples made the first attempt at sky-mapping; it may have been the Egyptians, the Chinese or even the Cretans. Certainly there were various different systems.

Naturally, the main emphasis was on observation, and in particular the Chinese records are most informative. There were good descriptions of unusual phenomena, such as eclipses and comets, and the dates are given accurately, though of course the Chinese had no idea of the causes of these events. We know that an eclipse of the Sun is due to the temporary covering-up of the brilliant disk by the dark, invisible body of the Moon; the Chinese preferred to think that the Sun was being attacked by a hungry dragon, so that the only remedy was to scare the beast away by making as much noise as possible. (It always worked! Sadly, the old story that two Court astronomers named Hi and Ho were executed for failing to predict an eclipse has been discounted by modern scholars.) The first Chinese record of a solar eclipse goes back to 2136 BC. Comets were also studied, and were regarded with alarm, because it was thought they indicated divine displeasure – a feeling which is not quite dead even now.

The Egyptians were avid sky-watchers, and there is no doubt that the Pyramids are astronomically aligned, but true astronomy began with the Greeks. It is often said that the 'Greek miracle' was rapid; actually it was nothing of the kind. Thales of Miletus, the first of the great philosophers, was born around 624 BC; Ptolemy of Alexandria, the last, died about AD 180 – giving a total span of over eight centuries, so that in time Ptolemy was as remote from Thales as we are from the Crusades. Yet it is true that the Greeks made amazing progress, and we can only admire them.

There were two great steps to be taken. One was to prove that the Earth is not flat, and the other was to show that it is moving around the Sun, so that it is not genuinely important and merely ranks as a normal planet. The first of these steps was taken at an early stage in the Greek story, and Aristotle, around 350 BC, knew quite well that the Earth is a globe. He was even able to give observational proof. First, he knew that the brilliant southern star Canopus can be seen from Alexandria, but not from the more northerly Athens; this is easy to explain if the world is spherical, but not if it is flat. Secondly, he knew that an eclipse of the Moon is caused by the shadow of the Earth falling on the lunar surface; since this shadow is curved, it follows that the Earth's surface must also be curved. Around 240 BC another philosopher, Eratosthenes of Cyrene, even made a remarkably accurate estimate of the Earth's size.

Later Greeks proved to be expert observers. Hipparchus of Nicaea, around 140 BC (all our BC dates are bound to be somewhat uncertain) drew up a good star catalogue, and the accuracy of his measurements is shown by the fact that he discovered the phenomenon of precession. Between AD 140 and 180 Ptolemy of Alexandria extended and improved Hipparchus' catalogue, and also produced the first map of the world which was based upon astronomical observation rather than guesswork. It was reasonably good, though naturally it covered only the Mediterranean and adjacent areas (Britain is shown, though Scotland is joined on to England at a curious angle). It was Ptolemy, too, who wrote a great book which has come down to us by way of its Arab translation, and is usually called the *Almagest*. It is really a compendium of ancient science, and its value to historians has been incalculable.

What the Greeks could not bring themselves to do was to dethrone the Earth from its proud central position. A few of the philosophers, notably Aristarchus of Samos around 280 BC, did take this drastic step, but as they could give no proof they found few followers, and the later philosophers went back to the idea of an all-important central Earth. This is always termed the Ptolemaic system, even though Ptolemy himself did not invent it, and only brought it to its highest degree of perfection (see Figure 3.1).

According to Ptolemy, the Earth lay in the centre of the universe, with the Moon, Sun, and planets and stars moving around it; the stars were carried on the outer crystal sphere, and other spheres carried the members of the Solar System. All orbits had to be circular, because the circle is the 'perfect' form, and nothing short of absolute perfection can be allowed in the heavens. The trouble about this idea was that the movements of the planets were not regular, and so could not be explained by uniform circular motion. Ptolemy, who was an excellent mathematician, knew this quite well, and was forced to adopt a cumbersome system with each planet moving around the Earth in a small circle or epicycle, the centre of which – the deferent – itself moved round the Earth in a perfect circle. In the end the whole plan became hopelessly involved and unwieldy, but it did fit the observations, and for many centuries it was not seriously challenged.

After Ptolemy's death the Dark Ages descended over Europe, and there was a period of stagnation, though there was an important development around AD 570 when Isidorus, Bishop of Seville, became the first major figure to draw a clear distinction between astronomy and astrology. The revival of astronomy was due to the Arabs, from the early ninth century; they drew up accurate star catalogues, and measured the movements of the Sun, Moon and planets more precisely than Ptolemy had been able to do, although admittedly their main aim was astrological. In 1433 Ulugh Beigh, one of the most powerful of the Oriental rulers, set up an elaborate observatory at his capital at Samarkand, and new, improved planetary tables were produced. But with Ulugh Beigh's murder the Arab school came to an end, and subsequent developments were chiefly European.

The next important step was taken in the sixteenth century by a Polish churchman whose name was Mikołaj Kopernik, known to us as Copernicus. He had made careful studies of the movements of the planets, and realized that the Ptolemaic theory was hopelessly artificial; most of the difficulties could be removed by the simple expedient of taking the Earth away from the centre of the system and putting the Sun there instead (see Figure 3.2). He was well aware that the Church would accuse him of heresy, and he withheld publication until the very end of

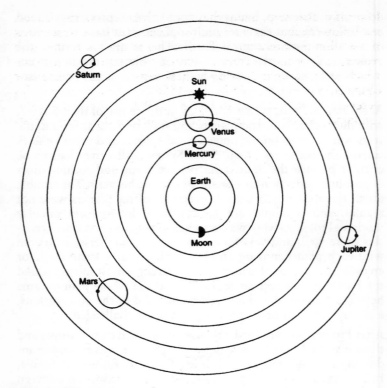

Figure 3.1 Ptolemaic system

his life, in 1543. This was a wise precaution, and the Church leaders lost no time in launching fierce attacks on him. Martin Luther referred to him as 'this fool who wants to turn the heavens upside-down', and persecution began; in 1600 one Copernican supporter, Giordano Bruno, was actually burned at the stake in Rome partly because of his expressed belief that the Earth moves around the Sun.

Adopting the Sun-centred or 'heliocentric' theory was Copernicus' only major achievement; most of his other conclusions were wrong, and he was wedded to the concept of perfectly circular orbits, so that he was even reduced to bringing back Ptolemy's epicycles. Yet he had taken the essential step, and his book, *De Revolutionibus Orbium Coelestium*, marked the beginning of the greatest of all revolutions in astronomical thought.

Ironically, the next main character in the story, the Danish nobleman Tycho Brahe, was no Copernican; he could not believe that the Earth could be relegated to the status of a mere planet, and he worked out a weird hybrid system of his own which satisfied nobody. However, he was a superbly accurate observer, and between 1576 and 1596 he worked away in his observatory at Hven, an island in the Baltic, compiling a star catalogue which was incomparably better than any of its

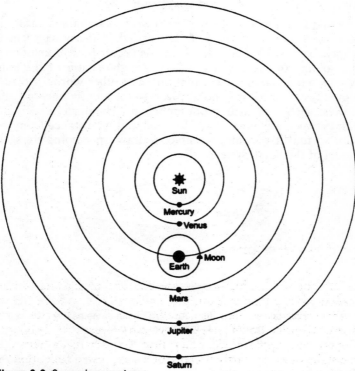

Figure 3.2 Copernican system

predecessors. He also made careful measurements of the movements of the planets, particularly Mars. When he died, in 1601, his work fell into the hands of his last assistant, a young German mathematician named Johannes Kepler, and the stage was set for the next major advance.

Tycho and Kepler were very different from each other. Tycho was haughty and ruthless, with a wonderful sense of his own importance (during his student days he lost part of his nose in a duel, and made himself a new one out of gold, silver and wax). Kepler was a frail, neurotic man, some of whose theories seem to belong to the mediaeval period rather than to the more modern age. But Kepler had implicit faith in Tycho's observations, and after years of work he found the answer to the main problem. The planets do indeed move around the Sun, but they do so in orbits which are elliptical rather than circular. Finally Kepler was able to draw up his three Laws of Planetary Motion, upon which all later work has been based (see Figure 3.3). The first two Laws were published in 1609, and the third in 1618.

The first Law states that a planet moves in an elliptical path; the Sun occupies one focus of the ellipse, while the other focus is empty. Law two states that the radius vector (i.e. an imaginary line joining the centre of the planet to the centre of the Sun) sweeps out equal areas in equal times; in other words, a planet moves fastest when it is at its closest to the Sun (perihelion) and slowest when it is furthest out (aphelion). The third Law gives a link between a planet's orbital period and its distance from the Sun, so that it becomes possible to draw up a complete scale model of the Solar System.

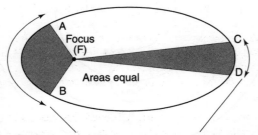

Time taken to move from AB to AC is the same as the
time taken to travel from AD to AC.
The area ABF is equal to the area FCD.

Figure 3.3 Kepler's Laws

It is worth noting here that most planetary orbits are not very different from circles; draw the Earth's path on a scale fitting this page, and it would be almost indistinguishable from a circle. Mars, upon which Kepler concentrated, has a rather more eccentric path, and the difference between its perihelion and aphelion is over 32 million km (20 million miles); however, Kepler's work was a brilliant piece of mathematical investigation. Meanwhile, telescopes had been invented.

It is widely believed that the first telescopes were made in Holland, in 1608. It is true that these are the first telescopes of whose existence we have conclusive proof, but research by the British historian C. A. Ronan now seems to have established that a curious kind of telescope was made by Leonard Digges, in England, at some time between 1550 and 1560, in which case he preceded the Dutch by over half a century. The Digges telescope appears to have used both mirrors and lenses; whether it was ever turned skyward we do not know, and neither have we an accurate idea of what it looked like. So astronomically, at least, telescopes come into the picture at around the time that Kepler was publishing his first Laws. Thomas Harriot, one-time tutor to Sir Walter Raleigh, drew a telescopic map of the Moon in 1609, but the first systematic observer was Galileo Galilei, the great Italian scientist who was also the true founder of experimental mechanics.

Galileo heard of the Dutch invention, copied and improved it, and used it to make a series of spectacular discoveries, confirming that Copernicus had been right in believing that the Earth moves around the Sun. Galileo saw the mountains and craters of the Moon; the countless stars of the Milky Way; the four bright satellites of Jupiter, proving at least that the Earth is not the only centre of motion in the universe; and, even more significantly, the phases of Venus. Because Venus is closer to the Sun than we are, it shows phases similar to those of the Moon, and for much the same reason. The point is that in Ptolemy's theory, Venus could never show a complete cycle of phases from new to full.

Unlike Kepler, Galileo was not cautious. He not only published his beliefs, but did so in a way which was tactless enough to offend the Holy See. He was accused of heresy, brought to trial by the Inquisition, and forced into a hollow and completely meaningless recantation. His major work, the *Dialogues*, remained on the Papal Index of forbidden books until 1836, and it was not until 1993 that the Vatican finally admitted that Galileo had been right all along!

The Ptolemaic theory was eventually abandoned in 1687, with the publication of Isaac Newton's immortal *Principia*, in which Newton laid down the laws of gravitation and ushered in the modern scientific era. This book has been described as the greatest mental effort ever made by one man, but it was by no means Newton's only achievement. It was he who used a glass prism to split up sunlight and show that it is made up of a combination of all colours of the rainbow, thereby introducing the science of spectroscopy; and in 1671 he presented the Royal Society with a telescope of an entirely new type.

Galileo's telescope was a refractor, collecting its light by means of a glass lens. One problem was that the lens produced false colour, so that an object such as a star was surrounded by gaudy rings which may have looked beautiful, but were most certainly unwanted. The initial remedy was to build telescopes of immense length, making them awkward to use; in extreme cases, the main lens had to be fixed to the top of a mast. Newton decided to do away with the object-glass altogether, and to collect the light by using a curved mirror, which reflects all the colours equally. Newton's first 'reflector' had a main mirror 2.5 cm (1 inch) across; today the world's largest telescope has a mirror almost 1.5 m (600 inches) in diameter. We have come a long way in a little over three centuries.

It was around this time that the first telescopically equipped observatories were established. So far as Britain was concerned, the main need was navigational, because there had been several naval disasters due to the fact that the commanders simply did not know where they were after spending weeks or months out at sea. Latitude-finding was easy enough; all that had to be done was to measure the altitude of the Pole Star, and then make a minor correction to allow for the fact that the Pole Star is not exactly at the pole. It was longitude-finding which was the main problem. Once the exact 'ship's time' was known and compared with the time at Greenwich, longitude could be worked out; but there were no clocks accurate enough to function properly at sea. The alternative was to use the changing position of the Moon as a sort of natural clock-hand, and measure its movements against the stars. This involved having an accurate star catalogue, and even Tycho's was not good enough because it had been compiled solely on the basis of naked-eye observations. (It is ironical that Tycho died just before the telescope became readily available; he would have made such good use of it!) Accordingly, that much-maligned monarch King Charles II ordered that a new observatory should be built and equipped with a telescope, so that the astronomer in charge could set to work in compiling a catalogue good enough to satisfy the sailors. The chosen site was Greenwich Park, in Outer London, and the Rev. John Flamsteed was appointed Astronomer Royal. He did eventually produce the required catalogue, though it took him a long time, and by the time it was finished the whole method of 'lunar distances' had become obsolete because of the development of accurate marine chronometers.

Flamsteed was succeeded by Edmond Halley, best remembered now because of his association with the comet which bears his name. Unfortunately the Royal Greenwich Observatory has now been closed down. The original building – designed, incidentally, by Sir Christopher Wren – has been turned into a museum.

Next we come to the career of Sir William Herschel, a Hanoverian musician who came to England while still a young man, and spent the rest of his life there. For a while he was organist at the fashionable resort of Bath Spa, but his main interest was astronomical, and he built his own reflecting telescopes. With one of these, in 1781, he discovered a new planet, now known to us as Uranus. The discovery made Herschel famous; a grant from King George III of England and Hanover made it possible for him to abandon music as a profession, and he devoted himself to 'reviews of the heavens', his main aim being to find out the way in which the stars are arranged in space. By then it was known that the stars are suns, and that they are very remote indeed; Herschel failed to measure their distances, but during his work he made countless important discoveries, so that he is often regarded as the greatest observer of all time. He came to the conclusion that the Galaxy must be a flattened system, and in this he was of course correct, though he erred in placing the Sun near the middle of the system. His largest telescope was a reflector with a 1.2-m (49-inch) mirror and a focal length of 12 m (40 ft), though most of his best work was carried out with smaller and more manageable instruments. Other observers were active at the same time – for example Charles Messier, who compiled the famous catalogue of star-clusters and nebulae, and Johann Hieronymus Schröter, who concentrated upon observations of the Moon and planets.

Up to the end of the eighteenth century almost all astronomers had been based in the Earth's northern hemisphere, so that the stars of the far south had been neglected. During the 1830s Sir John Herschel, son of William, took a telescope to the Cape of Good Hope and made the first detailed survey of the southern sky, adding to his father's amazingly numerous discoveries of objects such as double stars, clusters and nebulae. It was at this time, too, that photography began to play a significant role, and as the years passed by the camera began to supersede the human eye for most branches of observation. The change-over was virtually complete well before the end of the nineteenth century.

One fascinating development came from Ireland. At Birr Castle, in the middle of the island, the third Earl of Rosse built a monster telescope with a 1.8-m (72-inch) mirror, the largest ever made up to that time. It was clumsy and unwieldy, but it worked well, and Lord Rosse used it to great effect. He found that some of the nebulae seen by Herschel, Messier and others looked as though they were made of shining gas, but others were starry, and many of these were spiral in shape. Was it possible that these 'spiral nebulae' were external galaxies, rather than being mere minor features of the Milky Way? It was difficult to tell. In 1838 the German astronomer Friedrich Bessel had managed to make the first measurement of the distance of a star, and had shown that 61 Cygni, a dim star in the constellation of the Swan, lay at a distance of around 11 light-years, corresponding to 97 million million km (60 million million miles); but the spirals were clearly far more remote than this, and the problem of their status was not solved until 1923, but at least Herschel, Rosse and others had made the suggestion.

In 1666 Newton had broken sunlight into its constituent colours. It was not until a century and a half later that the German optician Josef Fraunhofer made the first close study of the nature of sunlight, and a full explanation had to wait until 1858, when two of Fraunhofer's countrymen, Gustav Kirchhoff and Robert Bunsen, showed that by using spectroscopes it was possible to find out 'what the stars are made of'. Today the spectroscope may be said to be the main research tool of the astronomer, and without it our knowledge of the universe would indeed be meagre.

Mathematical astronomy, too, had made great strides. In 1846 the new planet Neptune was tracked down because of the pull it had exerted upon Uranus, and this gave a final vindication of Newton's theories.

The Rosse 1.8-m (72-inch) reflector was soon overtaken. More powerful refractors were built; the largest of all, at the Yerkes Observatory in the United States, was given a 1-m (40-inch) lens. It was completed in 1897, and is not likely to be surpassed, because there is a limit to the size of a useful lens; if too large and heavy it will distort under its own weight, ruining its performance – remember that a lens has to be supported all around its edge. Reflectors took over, with mirrors made up of glass rather than metal as in the Rosse telescope. The lead was taken by George Ellery Hale, an American who not only planned large telescopes but also persuaded friendly millionaires to finance them. Hale set up a 1.5-m (60-inch) reflector on Mount Wilson in California, and followed this in 1917 with a 2.5-m (100-inch) reflector, which remained in a class of its own for more than three decades. It was with this telescope that Edwin Hubble was able to make the observations proving that the 'spiral nebulae' were, after all, independent galaxies far beyond our own. It was also Hubble who first demonstrated that the galaxies are racing away from us, and that the whole universe is expanding.

Mount Wilson is a lofty peak. Hale knew that the atmosphere is the main enemy of the astronomer; it is dirty and unsteady, and moreover it blocks out many of the important radiations coming from space. The solution is to 'go up' and this is why most great modern observatories are situated at high altitude – 4267 m (14 000 ft) in the case of the observatory on the summit of Mauna Kea, in Hawaii. There is also a tendency to concentrate upon the southern hemisphere, both because conditions there are generally better than those in the north and also because many of the most interesting objects in the sky are so far south that they never rise over Europe or mainland United States.

The major telescopes of today far outmatch those of even the recent past. The Mount Wilson 2.5-m (100-inch) was overtaken in 1948 by the 5-m (200-inch) reflector on Mount Palomar, also in California. It too had been master-minded by Hale, though sadly he did not live to see its completion. Yet by now even the 5-m (200-inch) is not in the 'top ten'. Mànne Kea is the home of several great reflectors: the 8.3-m (327-inch) Japanese Subaru, the 8-m (315-inch) Gemini North, and the twin Keck telescopes, each with a 10-m (387-inch) mirror. Working together the Kecks could, in theory, distinguish a car's headlights separately from a range of 25 750 km (16 000 miles). Ay Cerro Parañal, in the Atacama Desert of Northern Chile, hosts the VLT, or Very Large Telescope. This is actually made up of four telescopes (Antu, Kueyen, Melipal and Yepun), each with an 8-m (315-inch) mirror, which can work together. Other giants are being planned. We look forward to OWL, the Overwhelmingly Large Telescope, which is to have an equivalent aperture of no less than 100 m (328 ft).

Until the twentieth century, astronomers were limited to studying the rays of visible light coming to us from space, but in 1931 came the birth of radio astronomy. Light is a wave-motion, and the colour of the light depends upon its wavelength – that is to say, the distance between successive wave-crests. Red light has the longest wavelength and violet the shortest, though even for red light the wavelength is tiny by normal standards -- about 7500 Ångströms (one Ångström being equal to one hundred-millionth part of a centimetre; the name honours the nineteenth-century Swedish physicist Anders Ångström). But visible light is only a small part of the total range of wavelengths, or electromagnetic spectrum (see Figure 3.4). To the long-wave end of red we have infra-red, microwaves and then radio waves; to the short-wave end we have ultra-violet, X-rays and the very short gamma-rays. Very few of these 'invisible radiations' can reach ground level, because they are blocked out by layers in the Earth's upper air, but some can, and in particular there is a 'window' in the radio range. In 1931 Karl Jansky, a Czech-born American radio engineer, was carrying out investigations into static, using an improvised aerial, when he found that he was picking up radio emissions from the Milky Way. Curiously, he never followed up this discovery as he might have been expected to do, but after the end of the Second World War radio astronomy became a vitally important branch of science; Sir Bernard Lovell master-minded the huge 76-m (250-ft) radio 'dish' telescope at Jodrell Bank, in Cheshire, which came into operation in 1957, and by now there are many radio arrays all over the world. Of course, a radio telescope does not produce a visible picture in the same way as an optical telescope, and one certainly cannot look through it. It is more in the nature of a large aerial, but it can provide information which we could never obtain in any other way.

The Space Age began on 4 October 1957, with a very pronounced bang. The Russians launched the first of all artificial satellites, Sputnik 1, which was football-sized and carried little apart from a radio transmitter, but it ushered in a new era. It soon became possible to send sophisticated equipment above the atmosphere, and to study the radiations which can never reach ground level; for example X-ray astronomy began in 1962, when a rocket-borne detector tracked down the first discrete celestial X-ray source. The first probes to the Moon were dispatched in 1959; manned flight began two years later, with the epic journey of Yuri Gagarin, and before long manned journeys were common, culminating in the pioneer Moon landing by Neil Armstrong and Buzz Aldrin in July 1969. Meanwhile, automatic probes were being sent to the planets, and by now all the principal members of the Sun's family have been surveyed from close range. There have even been controlled landings on Mars, Venus, one asteroid (Eros), and Titan, the largest satellite of Saturn. It is probably true to say that we have learned more about the planets during the last 50 years than we had been able to do for the previous 40 centuries.

Optical telescopes have been improved beyond all recognition. However, all ground-based telescopes are handicapped by having to look through the atmosphere, and this is why a space telescope was planned. It was launched in 1990, and put into an orbit around the Earth over 563 km (350 miles) above ground level; it has a 2.4-m (94-inch) mirror, and 'seeing' conditions are perfect all the time, so that it can out-perform any telescope operating from the Earth's surface. Fittingly, it was named in honour of Edwin Hubble.

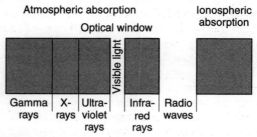

Figure 3.4 Electromagnetic spectrum

Just as photography superseded visual observation over a century ago, so photography itself is now being superseded by electronic equipment. CCDs – Charge-Coupled Devices – are far more sensitive than any photographic plate, and have largely taken over, which means that telescopes built years ago are now much more powerful than they originally were. No longer does an astronomer have to spend long hours in a cold, darkened dome, checking on his photographic exposures. Today, he sits in the comfort of a control room, watching a television screen. He need not be in the dome at all, or even in the same country. It is quite practicable for an astronomer in, say, Edinburgh to control a telescope in Hawaii.

I realize that this has been a very breathless and sketchy account of events over the ages, but I hope that it will help in setting the scene. Modern astronomy is a dynamic science; every year brings its quota of new discoveries and new surprises. We have learned a great deal, though we cannot yet pretend to have more than a vague understanding of the great universe in which we all live.

04

the astronomer's telescope

In this chapter you will learn:
- about the various types of astronomical telescopes
- how the telescopes work
- how to choose a telescope for yourself.

If you become seriously interested in astronomy, then sooner or later you will want to equip yourself with a telescope. This is often where the trouble begins because, unfortunately, there are some very poor telescopes on the market. I have always advised against spending too much money on a refracting telescope with a main lens less than 7.5 cm (3 inches) across, or a Newtonian reflector with a mirror smaller than 15 cm (6 inches). This is still sound advice, but I have to admit that there are now some small refractors on sale for less than £100 ($200) which are surprisingly good. If you want to make a really modest start, one of these will be a great deal better than nothing – but do not expect too much.

The alternative is to invest in a pair of binoculars, which have most of the advantages of a small telescope apart from sheer magnification, and which are relatively cheap. They can be used for everyday viewing, such as bird-watching, whereas a conventional astronomical telescope cannot, partly because of the small field of view and partly because it gives an upside-down image.

Binoculars are classified according to their magnification and the diameter of the main object-glass, always given in millimetres. Thus a 7 x 50 pair gives a power of 7, with each object-glass 50 mm (2 inches) across. This is suitable because a higher magnification – at least above x 12 – means that the binoculars become so unwieldy that they are difficult to hand-hold. (Remember not to look anywhere near the Sun. I will have more to say about this later, but I do not apologize for stressing it, because it is so important.)

The principle of the refractor is shown in Figure 4.1. Light from the object under study passes through the object-glass, and the rays are bunched up and brought to focus, where the image is magnified by a second lens known as an eyepiece or ocular. Note that it is the object-glass which is responsible for the light-collection, while all the actual magnification is carried out by the eyepiece – and if you see a telescope advertised as 'magnifying' so many times, without any mention of the size of the object-glass, avoid it completely. The distance between the object-glass and the focus is termed the focal length. Generally you will need at least three eyepieces, one giving low power for general views, one giving medium power for more detailed work, and one yielding high power for use on the Moon or planets under really good conditions.

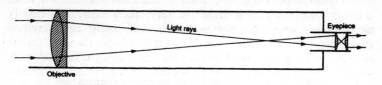

Figure 4.1 Refractor principle

The commonest form of reflecting telescope is the Newtonian (no prizes are offered for guessing who first worked out this system). The light passes down an open tube, which may be a skeleton, and falls upon a curved mirror at the far end (see Figure 4.2). This parabolic speculum sends the light back up the tube, directing it on to a smaller, flat mirror inclined at an angle of 45 degrees. The flat sends the rays into the side of the tube, where they are brought to focus and the image is enlarged by an eyepiece as before. With a Newtonian, then, the observer looks into the side of the tube instead of up it, and it is almost essential to have a finder telescope attached to the main tube.

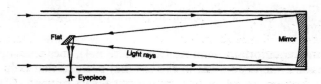

Figure 4.2 Newtonian reflector

Each type has its own advantages – and, unfortunately, its own drawbacks. Inch for inch, the refractor is the more effective of the two; it is also sturdier, and if well treated will not need drastic maintenance for many years, while reflectors are much more temperamental, quite apart from the fact that their mirrors have to be periodically recoated with either aluminium or silver. Against this, a Newtonian reflector is much cheaper than a refractor of comparable performance, and it does not produce the irritating false colour which so plagued the early observers (though it must be added that modern compound object-glasses reduce this trouble to a bare minimum). There are many other optical systems; for example there is the Cassegrain, where the secondary mirror is convex rather than flat, and the light is sent back to the eyepiece via a hole in the main speculum (see Figure 4.3). There are 'folded' telescopes which use both mirrors and lenses, and these are very convenient, particularly as they are easy to move around; their main drawback is their cost.

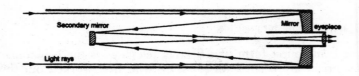

Figure 4.3 Cassegrain reflector

Mountings are of two main types. An altazimuth stand means that the telescope can be moved freely in any direction, either in altitude (up or down) or azimuth (east or west). The problem here is that two motions have to be considered, and guiding is not easy when a high magnification is being used; it is amazing how quickly a celestial object will shoot across the field of view! Neither can an altazimuth telescope be used for photography, except in a very limited sense. This also applies to the increasingly popular 'Dobsonian', which is elementary but, in its way, effective (see Figure 4.4).

With an equatorial mounting (see Figure 4.4), the telescope is set upon an axis pointing to the celestial pole, so that only one movement has to be taken into account – east to west. The up or down motion is automatic, and the addition of a mechanical drive means that the target object can be kept firmly in the field of view.

Setting up a new telescope can be a delicate and sometimes exasperating procedure. For example, you will need a finder – a very small telescope mounted on the main instrument. The finder will have a very wide field, and once the target object has been centred in the finder field it ought to be in the field of the main telescope as well. If not, then patient lining-up is essential, and this can be decidedly tricky. Once success has been achieved, make sure that the finder is firmly fixed!

Refractors are fairly straightforward, the main requirement being that the mounting must be steady. Some small commercial refractors are sold upon spidery tripods which quiver charmingly in the slightest breeze. Reflectors, however, can be more of a problem because they can easily go out of adjustment.

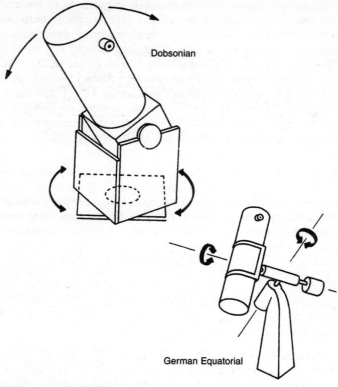

Dobsonian

German Equatorial

Figure 4.4 Mountings for telescopes

This is no place to go into detail about telescopic faults, but there are plenty of good instruction books, some of which are listed in the 'Taking it further' section.

Another problem is siting. Light pollution is now a serious menace to the astronomer – both amateur and professional. Some years ago the Mount Wilson 2.5-m (100-inch) reflector was temporarily mothballed because of the glare from the city of Los Angeles. Anyone living in a city will be lucky to see the stars at all, and the only real solution is to have a portable telescope. The best answer here is probably a Meade or Celestron, with an aperture of 10 cm (4 inches) or more; this is a 'compound' system using both lens and mirror. If you are prepared to spend, say, £500 ($1000), you can equip yourself well.

If you do have facilities for a permanent site, you may wish to buy or make an observatory. There are various patterns available. Of course a sleek, graceful rotating dome looks impressive, but it is not essential. One easy answer is a run-off shed, mounted on rails. In this case it is wise to make the shed in two parts, running back in opposite directions. A single shed must have a door, and this can pose problems. If hinged, it flaps. If it is removable and you try to replace it after an observing session on a dark, windy night, you may find that the door takes on the role of a powerful sail. A run-off roof structure is quite suitable for a refractor, but less so for a reflector.

Finally, make sure that you choose the best possible position for your observatory. It is usually the case that any adjacent trees lie in the most awkward places – and remember that trees have a nasty habit of growing. That innocent-looking sapling of today may turn into a monster, capable of obscuring a large part of the horizon.

All in all, it is sensible for the newcomer to obtain the best possible on-the-spot advice. There is bound to be a local astronomical society reasonably near at hand, and you will almost certainly find someone ready to help. Once you have acquired a good telescope, it will last you a lifetime, and the cost is non-recurring.

There is, of course, endless scope for astronomical photography, though if you are going to take photographs through the telescope you will need an equatorial mount and a clock drive. Otherwise, be content with taking pleasing pictures of star trails, using an ordinary fixed camera. With luck, you may even record a meteor or two. Displays of aurorae, or Northern Lights, are very common in latitudes such as those of Scotland, and can sometimes be bright even from South England; if one appears, do not miss the chance to photograph it. There is, too, the occasional spectacular comet, such as that of 2007. Limited though it may be, the humble hand-held camera is certainly not to be despised.

During the past few years the situations of photography has been transformed. Cameras, film or digital, are looking decidedly old-fashioned. Images obtained with modest telescopes, by means of CCDs (Charge-Coupled Devices) and similar electronic equipment, outmatch the best professional results of the late 1980s – and the cost is not great. We have entered the electronic age, and the amateur astronomer has become more valuable than ever before.

05

into space

The concept of space-travel is very old. People have long dreamed of breaking free from the Earth, and in the second century AD a Greek satirist, Lucian of Samosata, went so far as to write a story about a flight to the Moon. He called it the *True History*, because, in his own words, it was made up of nothing but lies from beginning to end. He explained how a ship passing through the Pillars of Hercules, now known to us as the Straits of Gibraltar, was caught up in a waterspout, and hurled upward so violently that after seven days and seven nights it reached the Moon – where the sailors found that they had arrived at a fascinating time; the King of the Moon was about to do battle with the King of the Sun with regard to who should have first claim on Venus (the planet, I hasten to add, not the goddess). Later ideas of propulsion involved bird-power, fire-crackers and even demons. Finally, in 1865, the first scientific method was described. It did not work, and could never do so, but it was of value none the less.

The author was Jules Verne, the great French story-teller. In his novel *From the Earth to the Moon* three adventurers climb inside a projectile, and are fired Moonward at a speed of 7 miles (11.2 km) per second. Verne believed in being as accurate as possible, and he did give the correct speed for departure; but there were several things that he either did not know or else chose to ignore. First, starting off at 7 miles (11.2 km) per second would be quite a jerk, to put it mildly, and would at once reduce the luckless travellers to jelly. Secondly, air sets up resistance, and resistance causes heat; even before it left the barrel of the gun, the projectile would have been burned away. And thirdly, the journey would be one-way only (though it is true that in his novel Verne cleverly avoided this particular difficulty).

Why 7 miles per second? This is the Earth's escape velocity (see Figure 5.1). Throw an object upward, and it will rise to a certain height before falling back. Throw it faster, and it will rise higher. At an initial speed of 7 miles (11.2 km) per second (approximately 40 225 kmh/25 000 mph) it would never return at all; the Earth's gravitational pull would not be strong enough to draw it back, and the object would escape into space. (The escape velocity of a body depends upon its mass; for example the Moon has only 1/81 of the mass of the Earth, and the lunar escape velocity is no more than 2.4 km (1.5 miles) per second.)

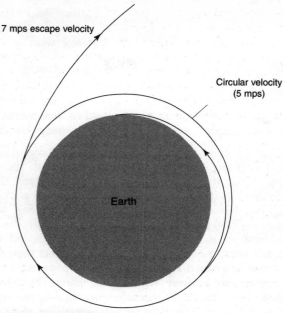

Figure 5.1 Escape velocity diagram

Space-guns being rejected, we come back to the only device which will operate in airless space: the rocket. The principle is exactly the same as with the firework rockets fired on 5 November each year in memory of Guy Fawkes (who, some people claim, is the one man we now need in Parliament). The firework consists of a hollow tube filled with gunpowder, plus a stick to add stability. When the touch-paper is lit, the gunpowder starts to burn; gas is shot out of the exhaust, and 'kicks' the tube in the opposite direction. So long as the gas continues to stream out, the rocket will continue to fly. Atmosphere is not needed, and is actually a nuisance, because it sets up resistance and has to be pushed out of the way. We depend upon what Isaac Newton called the principle of reaction – 'every action has an equal and opposite reaction' (see Figure 5.2).

There is a good way to demonstrate what is meant. Put a book on the floor, stand on the book and jump off. You will go one way, and the book will go in the opposite direction, because you have kicked against it; the effect would be the same if you were

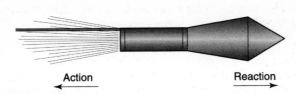

Action   Reaction

Figure 5.2 Rocket principle

in airless space (assuming that you could survive there). So if we want to invade space, rockets must be our means of travel.

The first truly scientific suggestion was made a century ago by a shy, deaf Russian schoolteacher who rejoiced in the name of Konstantin Eduardovich Tsiolkovskii. He realized that solid fuels are inadequate, and so he proposed a proper rocket motor, in which two liquids – say petrol and liquid oxygen – are forced into a combustion chamber, where they react to produce the gas sent out from the exhaust. Tsiolkovskii also suggested mounting rockets one on top of the other, so that in effect the uppermost stage would be given a running start into space. At the time his work caused no interest whatsoever, partly because it was published in an obscure Russian journal; but by the time of his death, in 1934, he had become recognized as the 'father of space-flight', even though he was not a practical experimenter and never personally fired a rocket in his life.

The first liquid-fuel rocket was sent up in 1926 by an American, Robert Hutchings Goddard, who had never heard of Tsiolkovskii's work. His rocket was modest enough, rising to a height of some metres (a few tens of feet) at a maximum speed of 97 kmh (60 mph), but it showed that the principle was valid. A few years later a German team set up an experimental research station near Berlin and began to build bigger and better rockets, but before long the politicians stepped in. The German Government swooped on the rocketeers and transferred its leading members, including Wernher von Braun, to Peenemünde, an island in the Baltic, with orders to produce weapons for the coming war. It was here that von Braun and his colleagues produced the V2 weapons used to bombard England during 1944 and 1945. Of course they were purely military, but they were rockets none the less, and they were the direct ancestors of the space-ships of today. After the German collapse, the experimenters were transferred to America, and by 1949 the upper stage of a purely scientific rocket had soared to the dizzy height of almost 402 km (250 miles). There was great activity at Cape Canaveral in Florida, the main rocket base, and by 1955 the White House felt confident enough to state that an artificial satellite would be launched some time during 1957 or 1958.

However, the Americans were forestalled. Russia's Sputnik 1 opened the Space Age on 4 October 1957; it entered a closed orbit around the Earth, sending back the never-to-be-forgotten 'Bleep! bleep!' signals. Other Russian satellites followed, and then, in 1958, von Braun master-minded the first American artificial satellite, Explorer 1. The 'race into space' was under way.

If an object is put into orbit with a velocity of 8 km (5 miles) per second, it will not fall down, any more than the Moon does (see Figure 5.3); it will continue circling the world indefinitely – unless any part of its orbit brings it down into the denser part of the atmosphere, so that it will be braked by friction and will eventually spiral to destruction (as happened to Sputnik 1 in early January 1958). Once the breakthrough had been achieved, satellites of all kinds were launched. Some carried scientific instruments to study radiations inaccessible from ground level (X-rays, for example); some were used for radio and television relays; others, regrettably, for military purposes.

Four years after Sputnik 1, the first man went into space. Predictably, he was a Russian – Yuri Gagarin – and he completed a full circuit of the Earth in his cramped capsule, Vostok 1, before landing safely in the prearranged part of what was then the Soviet Union. Up to that time it had been feared that space would be a hopelessly hostile environment. For example, there was potential danger from cosmic rays, which are not rays at all, but high-speed atomic particles bombarding the Earth from all directions all the time, though the 'heavy' particles are broken up in the high atmosphere and reduced to harmless fragments. Meteoroids were another cause of anxiety. But above all there was the question of weightlessness, or zero gravity. How would the human body react?

When in orbit around the Earth – in what is termed free fall – an astronaut has no apparent weight, not because he has 'got out of gravity' (which is theoretically impossible) but because he is neutralizing it. To show what is meant, put a coin on top of a book and hold it out at arm's length. The coin is pressing on the book, so that with reference to the book the coin is 'heavy'. Now drop the book. During the descent to the floor, the two are moving in the same direction at the same rate; the coin ceases to press on the book, so that with reference to the book it has

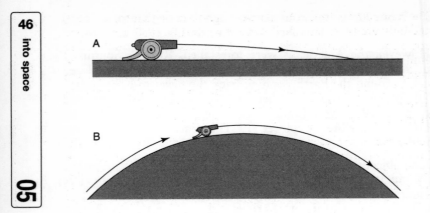

Figure 5.3 Orbital speed. Diagram A shows ballistic trajectory; diagram B shows orbital trajectory

become weightless (the effect would be just the same if the two were moving up, not down). Similarly, an orbiting astronaut will cease to press on his rocket, and will be weightless. Fortunately the effects have turned out not to be harmful, at least in the short term, and I remember Gagarin himself telling me that he found zero gravity 'very pleasant'! It was sad that Gagarin was killed in an ordinary aircraft crash a few years later.

Before long, two-person and even more elaborate vehicles were sent up, and women joined in (Valentina Tereshkova, in 1963, was the first). By the 1970s genuine space-stations had been launched, beginning with America's Skylab, which was manned by three successive crews and carried telescopes and scientific equipment of all kinds. The Russian space-station Mir was launched in 1986 and remained in orbit until 2001. Despite the various problems during the latter part of its career, it was a tremendous success and paved the way for the International Space-Station which is now circling the Earth.

One obvious need was for a ferry-vehicle, capable of taking crews and supplies up and down to space-stations. Using a new vehicle for every journey would be prohibitively expensive, and rather like using a new train for every journey from London to Bristol. This was the thinking behind the Space Shuttle, which is now an essential part of the whole space programme.

The first rockets to the Moon were Russian, beginning in 1959, and only ten years later Neil Armstrong and Buzz Aldrin stepped out on to the barren rocks of the lunar Sea of Tranquillity. No humans have been to the Moon since 1972, but there will be many more journeys there during the present century, and the idea of a fully fledged Lunar Base seems much less futuristic now than it did only a few years ago. But the Moon is our companion in space; what about journeys to the planets?

Sending a probe to a planet is much more difficult than to the Moon. The Moon is our companion, and always stays close to us, but even Venus, the nearest of the planets, is always at least a hundred times as distant, and it orbits the Sun rather than the Earth. Moreover, it is not possible to wait until it is at its closest to us, and then simply fire a rocket across the gap; this would mean using fuel for the entire journey, and no space-craft could possibly carry enough. What we have to do is to make use of the Sun's gravity, and put the probe into what is termed a transfer orbit, so that it can 'coast' unpowered for most of the way (see Figure 5.4).

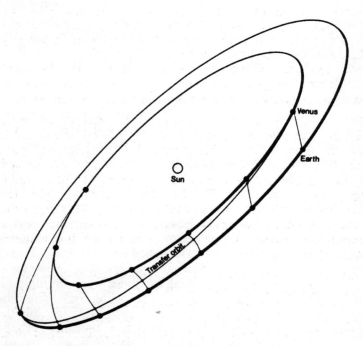

Figure 5.4 Transfer orbit

With an inner planet (i.e. Mercury or Venus) the probe has to be slowed down relative to the Earth. It will then start to swing inward towards the Sun, and the aim is to make it reach the orbit of the target planet at exactly the right moment. Mariner 2, the first of all successful interplanetary vehicles, was launched on 27 August 1962, and made its rendezvous with Venus on the following 14 December. (In case you are wondering what happened to Mariner 1, I have to tell you that it fell into the sea immediately after take-off; someone had forgotten to feed a minus sign into a computer.) Mariner 2 did not land, but simply by-passed Venus at a range of 33 790 km (21 000 miles), sending back valuable information before going on in a never-ending journey around the Sun; contact with it was finally lost in January 1963. Since then there have been many Venus probes, and the Russians have even made controlled landings there, obtaining pictures from the surface.

Mercury, the other inner planet, was first visited by Mariner 10, which made three active passes in 1974 and 1975 and is still, no doubt, orbiting the Sun. Mariner 10 was the first probe to make use of what is termed the gravity-assist technique, since it passed Venus *en route* and used Venus' gravitational pull to put it into an orbit which took it on to a rendezvous with Mercury. The next Mercury probe, Messenger, made its first close fly-by of Mercury in 2008, preparatory to entering a closed orbit round the planet.

With a superior planet – that is to say, a planet further away from the Sun than we are – the procedure is to speed up the space-craft relative to the Earth, so that it will swing outward. The first successful mission to Mars was Mariner 4, launched by the Americans in November 1964, which by-passed Mars in the following July. Others have followed and, due to controlled landings, we have been able to obtain information direct from the Martian surface.

Next came the outer planets. Naturally, Jupiter was the first target, with Pioneer 10, launched in March 1972; it made its pass of the Giant Planet in December of the following year. Pioneer 11 was an almost identical vehicle, sent up in 1973 and arriving in 1974, but after making its rendezvous with Jupiter it retained enough propellant for it to be swung back across the Solar System to an encounter with Saturn in 1979. But the Pioneers, good though they were, were upstaged by the Voyagers, both of which functioned flawlessly and are still sending back signals.

Voyager 1 was dispatched in 1977, and passed Jupiter in 1979. It then used the gravity-assist technique to go on to an encounter with Saturn, after which it began a never-ending journey out of the Solar System altogether. Its twin, Voyager 2, was even more ambitious (see Figure 5.5). By a fortunate chance, during the late 1970s, the four giant planets Jupiter, Saturn, Uranus and Neptune were strung out in a long, gentle curve, so that Voyager 2 was able to pass by them in turn – Jupiter in 1979, Saturn in 1981, Uranus in 1986 and Neptune in 1989. By the time it made its rendezvous with Neptune it had been in space for over 12 years and had covered more than 7240 million km (4500 million miles) – and yet it made its closest approach within two minutes of the scheduled time.

The Galileo probe to Jupiter was launched in 1989 but did not reach its target until 1995. During its journey it made fly-by encounters with Venus and the Earth, and sent back images of two asteroids, Caspra and Ida. Galileo entered orbit around Jupiter and began a detailed survey of the planet and its satellites. It also dispatched a separate probe, which impacted Jupiter and caused disturbances in the Jovian clouds which persisted for months. The Cassini probe to Saturn was launched in 1997 and reached its target in 2004; it carried a lander, Huygens, which came down successfully on Titan and transmitted direct from the surface. This was possibly the most difficult space operation yet attempted.

This is no place to go into details of the space-craft themselves; it is enough to say that they were immensely complex. Although it is possible to use solar power in our part of the Solar System, this cannot be done at remote depths, because there is not enough sunlight; the Voyagers carried tiny nuclear generators, and they worked well. Of course, the amount of power sent back to Earth was minuscule, and would not have been enough to light up the bulb of a pocket torch for more than a fraction of a second, but electronic techniques can use this tiny pulse to produce the superb pictures which have told us so much about the outer planets.

Comets, too, have been investigated. In 1986 the Giotto space-craft lunged into Halley's Comet and obtained the first ever images of a cometary nucleus from close range. In September 2001 the probe Deep Space 1 encountered Borrelly's periodical comet and excellent pictures were sent back.

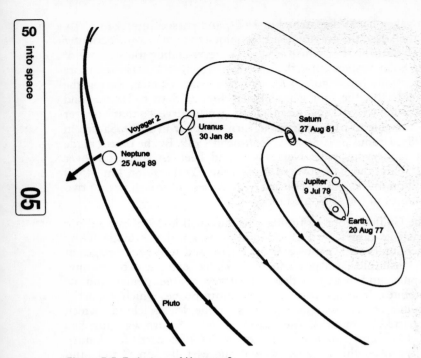

Voyager 2

Uranus
30 Jan 86

Saturn
27 Aug 81

Neptune
25 Aug 89

Jupiter
9 Jul 79

Earth
20 Aug 77

Pluto

Figure 5.5 Trajectory of Voyager 2

This is a book about astronomy, not space research, but the two are now so closely intertwined that it is impossible to separate them, any more than one can separate arithmetic from algebra. I felt therefore that it was essential to say at least a little about the events following 4 October 1957; now let us return to the sky itself.

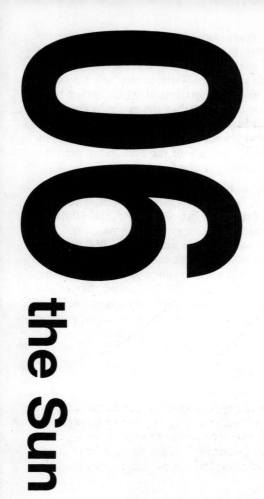

06

the Sun

In this chapter you will learn:
- how the Sun's distance was measured
- about the Sun itself and how it produces its energy
- about solar eclipses.

More than a hundred years ago a famous popularizer of astronomy, R. A. Proctor, wrote a book about the Sun. He subtitled it 'Ruler, Fire, Light and Life of the Planetary System', and with this nobody can disagree. Without the Sun we could not exist for a moment, and the Earth itself would never have come into existence. Yet it is only within the last two centuries that we have started to understand its nature, and there are still many important points about which we remain mystified. It has even been said that we know less about the Sun today than we thought we did at the beginning of the Space Age!

The Sun is, of course, quite unlike the Earth or any planet. It is made of incandescent gas, and it is extremely hot. Even the visible surface is at a temperature of between 5000° and 6000° C, and at the core, where the solar energy is being produced, the temperature soars to the unbelievable value of 15 000 000° C. The diameter of the globe is 1.4 million km (865 000 miles), so that the Sun could swallow up over a million globes the volume of the Earth, but the mass is 'only' 330 000 times that of the Earth, because the density is much lower (see Figure 6.1). In fact, the Sun 'weighs' only 1.4 times as much as an equal volume of water would do.

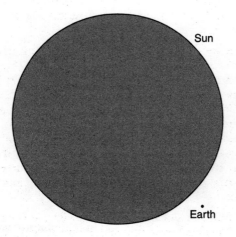

Figure 6.1 Sun and Earth compared

Attempts to find the distance between the Earth and the Sun were made in very early times, and although all these first efforts were wide of the mark they were at least able to show that the Sun is a long way away. The best of the Greek estimates was that of Ptolemy, who put the Sun's distance at 8 million km (5 million miles), but the first reliable measurement was made in 1672 by the Italian astronomer G. D. Cassini, who was at that time working in France as Director of the new Paris Observatory. Cassini's method involved using Kepler's Laws.

Johannes Kepler, remember, was the first to show that the planets move around the Sun in elliptical paths. His third Law gave a definite relationship between the orbital period and the distance from the Sun; for instance, since he knew the orbital period of Mars (687 Earth days) and that of the Earth (in round figures 365 days) it was simple to show that Mars must be 1.52 times further away from the Sun than we are. Very well, then: find the actual distance of Mars, in km/miles, and that of the Earth can be found by straightforward mathematics. Therefore Cassini set out to find the distance of Mars, and he used the method of parallax, measuring the planet's position among the stars from sites on the Earth's surface as widely separated as possible. His final result for the Earth–Sun distance, or astronomical unit, was 138 million km (86 million miles). Later measurements increased this to nearly 150 million km) (93 million miles), and now we can use an entirely different method – that of radar.

Radar involves sending out a pulse of energy, bouncing it off a solid body (or equivalent) and then receiving the 'echo' or return signal. Radar pulses have been bounced off the planet Venus; we know how fast they travel, because their speed is the same as that of light (299 270 km/186 000 miles per second), and so the time taken for the pulse to travel from the Earth to Venus and back again gives us the distance of Venus. From this, the length of the astronomical unit can be worked out. It proves to be 150 million km (92 976 000 miles) (this is of course the mean distance, because the Earth's path around the Sun is not a perfect circle).

We know the Sun's distance; we know its size, and we know its temperature. We also know that its age must be rather greater than that of the Earth, which is around 4600 million years. So how exactly does the Sun shine?

Some of the ideas current in the early and even mid-nineteenth century were rather curious. Sir William Herschel, discoverer of

Uranus, firmly believed that the brilliant solar surface lay above a calm, pleasant region which might well be inhabited – and he remained convinced of this throughout his lifetime (he died in 1822). Few people agreed with him, and it was tacitly assumed that the Sun must be burning in the manner of a vast bonfire. Unfortunately, rudimentary calculations showed that this could not be so. A Sun made up of coal, burning as fiercely as the real Sun actually does, would be reduced to ashes in less than a few thousand years. Another idea was that the Sun's energy was due to constant peppering with particles coming from space, but this was found to be equally inadequate. The contraction theory proposed by H. von Helmholtz in 1834 was rather better; this time it was assumed that the Sun is shrinking by about 60 m (200 ft) per year, releasing gravitational energy in the process. This would keep the Sun radiating for at least 15 million years – but this again was not nearly long enough. The real solution came much later, and was based upon that all-important tool of the astronomer, the spectroscope.

We have seen that, just as a telescope collects light, so a spectroscope splits it up. Pass a beam of sunlight through a glass prism, as Isaac Newton did in 1666, and it will produce a rainbow band, from red at one end through orange, yellow, green and blue to violet at the other (see Figure 6.2). This is because the prism bends or refracts the various colours differently: red least, violet most. It was in fact Newton who produced the first solar spectrum, though he never took his investigations much further.

The next step came in 1802 with the work of an Englishman, W. H. Wollaston. He passed sunlight through a slit, analysed it with a prism and saw that the rainbow band was crossed by

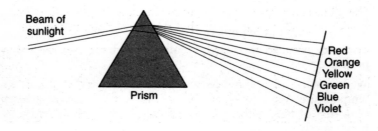

Figure 6.2 Production of a spectrum using a prism

dark lines. It was then that he missed scientific immortality. He thought that the lines merely marked the boundaries between the different colours, and did no more.

Next came a brilliant young German optician, Josef Fraunhofer, who had a romantic career; he was apprenticed to a Dickensian taskmaster, but when his ramshackle dwelling collapsed he was rescued and befriended by the Elector of Bavaria, who happened to be driving past in his coach. Subsequently Fraunhofer became the leader in his field, and it was he who, in 1814, went back to studies of the solar spectrum. Like Wollaston, he saw the dark lines. Unlike Wollaston, he realized that they were immensely significant. They always kept to the same positions, with the same intensities; he catalogued over 500 of them, and even today they are still often known as the Fraunhofer lines (see Figure 6.3).

Unfortunately Fraunhofer died young, at the very height of his powers, and the dark lines were not explained until 1859, as a result of work mainly by Gustav Kirchhoff. Kirchhoff found that spectra are of two different types. An incandescent solid, liquid, or gas at high pressure will produce a continuous rainbow band from red at the long-wave end through to violet at the short-wave end. An incandescent gas at low pressure will yield disconnected bright lines, each of which is the trademark of some particular element or group of elements; for example, two prominent yellow lines in one position in the spectrum must be due to sodium. See those yellow lines, and you know that sodium is present.

The Sun's bright surface or photosphere produces a continuous spectrum. Above it is a region of much more rarefied gas, termed the chromosphere, which should yield a bright-line or emission spectrum; but when seen against the rainbow background, the lines are 'reversed', and appear dark. This does not affect their positions or their intensities. Look at the yellow part of the band, and you will see two conspicuous dark lines. These lines correspond exactly to the familiar bright yellow lines of sodium, and we can prove that there is sodium in the Sun.

Absorption or Fraunhofer lines

Figure 6.3 Solar spectrum

Some elements produce very complicated spectra (thousands of lines in the case of iron, for example), but by now all the 92 naturally-occurring elements have been tracked down in the Sun. Helium, the second lightest of all the elements, was found in the solar spectrum years before it was isolated on Earth, and this accounts for its name – *helios* is Greek for 'Sun'. The English astronomer Norman Lockyer found it in the Sun in 1868, and not until 1894 was it identified on Earth.

All in all, we can now tell 'what the Sun is made of'. It is interesting to recall that in 1830 the French philosopher Auguste Comte stated that the chemistry of the stars was something which mankind could never find out.

The main constituent of the Sun is hydrogen, and this is no surprise, because hydrogen is the commonest element in the entire universe; indeed, atoms of hydrogen outnumber the atoms of all other elements put together. For every million hydrogen atoms in the Sun there are 63 000 of helium; next comes oxygen, with less than 700. So hydrogen accounts for 71 per cent of the Sun by mass, helium 27 per cent, and all the other elements combined less than 2 per cent.

Now we can start to investigate the way in which the Sun produces its energy; we have to deal with nuclear reactions near the core, where the temperatures and pressures are so high. For a while it was thought that the energy was due to atoms literally annihilating each other, but on this theory the Sun could last for millions of millions of years, giving a timescale which was as obviously too long as the contraction-theory timescale had been too short. The basic process was worked out in 1938 almost simultaneously by two scientists, Hans Bethe in America and George Gamow in Europe. Bethe had been attending a conference in Washington, and was returning to his university by train when he started to think about the solar energy problem; by the time he drew into his home station he had broken the back of the main puzzle!

Hydrogen is the key, and acts as the Sun's 'fuel'. At the core, the nuclei of hydrogen atoms are running together to make up the nuclei of the next lightest element, helium. It takes four hydrogen nuclei to produce one nucleus of helium, and every time this happens a little energy is set free and a little mass is lost, because the newly-created helium nucleus 'weighs' slightly less than the original four hydrogen nuclei combined. It is this released energy which keeps the Sun shining, and the loss in mass amounts to 4 million tonnes per second. This sounds alarming, but the Sun is so massive that by solar standards the loss is trifling, and there is enough 'fuel' to maintain the Sun in more or less its present state for several thousands of millions of years yet.

Of course, the supply of hydrogen will not last indefinitely, and in the long run it will become exhausted. When this happens, the Sun will have to change its structure. The core will shrink and become even hotter, while the outer layers will swell out as different reactions begin. The Sun will become a red giant star, and this is certain to mean the end of the Earth, so that our world has a limited lifetime, and if humankind still survives in that far-off period it will be time to consider migrating to a safer home! The Sun will then shed its outer layers altogether, and the remaining core will become very small and amazingly dense, glowing feebly as a 'white dwarf' star before losing the last of its light and heat and ending up as a cold, dead globe. I will have more to say about stellar evolution later, but at least we may be sure that nothing drastic is likely to happen yet.

On the other hand, the Sun is variable to some extent, and this is shown by the behaviour of the dark patches known as sunspots.

The bright solar surface, or photosphere, is generally disturbed by one or more of these patches. They have been known ever since they were studied with telescopes, from 1609 onward, and occasional spots visible with the naked eye were recorded from Ancient China, but of course their nature was unknown, and it was often supposed that they were openings in the bright photosphere, revealing darker, cooler material beneath. However, we now know that they are essentially magnetic phenomena. The Sun does have a magnetic field, and the lines of force run from one magnetic pole to the other, not very far below the surface of the photosphere. When the lines break through to the surface itself they temporarily cool it, and a sunspot results. In most cases a spot-group consists of two main outbreaks, which, predictably, have opposite magnetic polarity.

The Sun rotates on its axis – not as a solid body would do, because the equatorial region rotates faster than the poles; the equatorial period is 25 days, the polar period 34 days (this sort of behaviour is known as differential rotation). A sunspot will be carried slowly across the Earth-turned disk, and eventually it will vanish over the limb, to reappear a fortnight or so later, assuming that it has not disappeared in the meantime. No spot-group lasts for very long. A major group may persist for some months, but single small spots may have lifetimes of only a few hours.

A century and half ago, a German apothecary named Heinrich Schwabe was making daily observations of the Sun when he discovered something very interesting. There is a roughly regular solar cycle (see Figure 6.4). Every 11 years the Sun is active, with many spots and spot-groups; activity then dies down, and at minimum there may be many consecutive days with no spots at all, after which the groups become more common again up to the time of the next maximum. Thus there were maxima in 1958, 1969, 1991 and 2001; in the mid-1980s and mid-2000s activity was low by solar standards.

It was then found that there is another law, this time relating to the latitudes of the spots. The first spots of a new cycle appear at fairly high solar latitudes; later outbreaks appear closer to the Sun's equator, though they never actually break out on the equator itself. Even before the last low-latitude spots of the old cycle die out, we see the first spots of the new cycle appearing at higher latitudes. This is known as Spörer's Law, after the German who was one of the first to draw attention to it. The polarities of two-spot groups are also affected by the cycle. Suppose that in the Sun's northern hemisphere the 'leader' is a north-polarity spot and the 'follower' south-polarity; this will apply to all northern-hemisphere groups, while in the southern hemisphere it is the leader which will have south-polarity. At the end of the cycle there is a full reversal, so that it may even be said that the true length of a cycle is not 11 years, but 22.

The cycle is not perfectly regular, and the 11-year interval between one maximum and the next is merely an average. Moreover, it seems that between 1645 and 1715 there were almost no spots at all, so that the cycle was suspended; this is known as the Maunder Minimum, because the English astronomer E. W. Maunder was one of the first to note it by examination of the old records. We cannot be sure that spots were completely absent, because the records are so incomplete, but it does seem that they were rare, and there is also some evidence of earlier spot-free periods.

The Earth's climate is affected by solar activity, and fewer spots means colder spells, for example. It may or may not be significant that the Maunder Minimum coincided with what has been called the Little Ice Age in Europe; in England, for example, the River Thames froze almost every winter during the 1680s, and frost fairs were held on it, while there was serious talk of evacuating Iceland altogether. But it would be idle to pretend that we really know all we would like to know about the Sun.

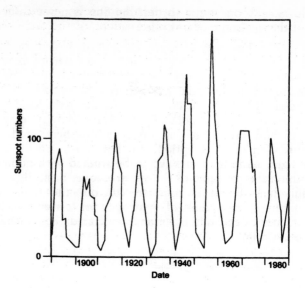

Figure 6.4 Solar cycle. Sunspot numbers over 100-year period, 1880–1980

A large spot consists of a dark central portion or umbra, surrounded by a lighter area or penumbra. (In fact, no spot is really dark; it appears so only because it is around 2000° C cooler than the surrounding photosphere. If it could be seen shining on its own, it would be intensely brilliant.) Some spots are regular in outline, but others are not, and there may be many umbrae contained in a single mass of penumbra, while single spots are also common. When a spot nears the Sun's limb it appears foreshortened, and from this it seems that most spots – not all – are depressions; the penumbra is always narrowest towards the limb, as was pointed out by the Scottish astronomer A. Wilson as long ago as 1774. Not all spots show the Wilson effect, but it is always worth looking for (see Figure 6.5).

Other features of the photosphere include *faculae* (Latin, 'torches'), which are bright areas usually associated with spot-groups, and are made up of high-level incandescent gases; they are best seen near the limb, where the photosphere is less brilliant than at the centre of the disk, and spectroscope analysis shows that they are made up of hydrogen. The surface also has a granular structure; each granule is around 9650 km (6000 miles) across, and has a lifetime of about eight minutes. It has been estimated that the entire surface contains about 4 million granules at any one time, so that the Sun is never calm even when at the minimum of its cycle.

Active spot-groups may produce short-lived, violent outbreaks known as flares, which send out electrified particles as well as short-wave radiations; it is these particles which cross the 150-million km (93-million-mile) gap between the Sun and the Earth and cascade down into the upper air, producing the lovely glows known as aurorae or polar lights. Flare activity may also interfere with layers in our upper atmosphere, and cause problems with radio reception. But few flares can be seen in visible light, so that most have to be studied with equipment based upon the principle of the spectroscope. Basically, a filter or equivalent device is used to cut out the light of all wavelengths except that of hydrogen or another selected element, such as calcium; flares and associated phenomena then become striking.

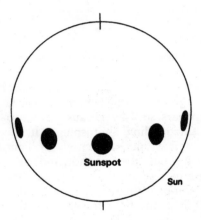

Figure 6.5 Wilson effect (the spot size is deliberately exaggerated)

Observing the Sun is a fascinating pastime, but it can be very dangerous indeed, and there are traps for the unwary. Remember, the Sun is very hot. Look straight at it through any telescope or pair of binoculars, and you will focus all the light and heat on to your eye, with tragic results. Using a dark filter in this way is emphatically not to be recommended, because the filter is always liable to shatter without warning, and in any case will not give full protection. It is sad but true that many people have blinded themselves permanently by using telescopes to look at the Sun, and I have known two cases myself. The only sensible method is to point the telescope at the Sun without looking through it, and then project the solar image on to a white screen, held or fixed behind the eyepiece (see Figure 6.6). True, there are some 'wedges' which are harmless enough, but my advice is never to use them unless you have expert guidance. In any case, projecting the image gives satisfactory results, and it is easy to make daily plots of the groups as they cross the disk from one limb to the other. You can also work out what is called the daily Zürich number, calculated by the formula $R = k (10g + f)$, where R is the Zürich number, g is the number of separate groups on view, and f is the number of separate umbrae. k is a constant depending upon the observer's equipment and experience, but it is usually close to unity. These are also known as Wolf numbers, as the system was devised by R. Wolf of Zürich in 1852.

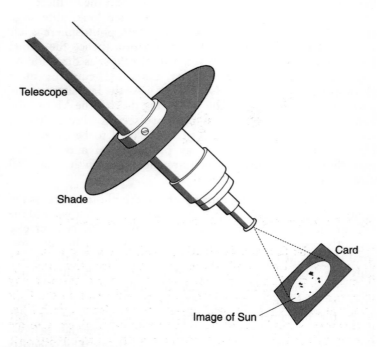

Figure 6.6 Projecting the Sun

Above the photosphere we have the chromosphere or 'coloursphere', and beyond this again comes the Sun's outer atmosphere, called the corona. Spectroscopic equipment can show the chromosphere and associated phenomena at any time, but for a good view of the corona we must either take a trip into space or else wait for a total solar eclipse.

During its monthly orbit around the Earth, there must obviously be times when the Moon passes directly between the Earth and the Sun. It so happens that the Sun and the Moon appear almost exactly the same size in the sky; the Sun's diameter is 400 times that of the Moon, but it is also 400 times further away. This is sheer coincidence – it can be nothing else – but it is fortunate for us.

Solar eclipses are of three types: partial, total and annular (see Figure 6.7). During a partial eclipse only a limited area of the Sun is hidden and although it makes for an interesting spectacle it is not particularly striking. Neither is an annular eclipse, when the lining-up is exact but the Moon is at or near the furthest part of its orbit around the Earth; it does not then look quite large enough to cover the Sun, so that a ring of the photosphere is left showing around the dark disk of the Moon (hence the name: Latin *annulus*, a ring). However, a total eclipse is dramatic by any standards, and may well lay claim to being the most glorious sight in all nature (see Plate 1). As the last sliver of the photosphere is hidden by the advancing disk of the Moon, the Sun's atmosphere flashes into sight; we see the chromosphere, with the red masses of hydrogen which used to be called Red Flames but are now known as prominences, and also the marvellous pearly corona, which stretches out from the Sun in all directions.

The chromosphere, lying directly above the photosphere, is the region where the Fraunhofer lines are produced; in it we find the prominences, which may be either relatively quiescent or else violently eruptive. They are not always present, but at most total eclipses at least one or two are seen. (Incidentally, it was only after the eclipse of 1852 that astronomers finally became certain that they were solar rather than lunar, which shows how little we knew about the Sun 160 years ago!) Because we can study the prominences with spectroscopic equipment at any time, without waiting for an eclipse, we know how they behave; but the corona is much more elusive, and this is why scientists are still willing to go on long journeys to make the most of a few fleeting seconds of totality.

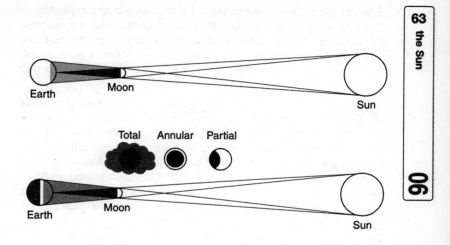

Figure 6.7 Theory of a solar eclipse

The corona is made up of gas, chiefly hydrogen. Its density is very low – less than one million-millionth of that of the Earth's air at sea-level – but its temperature is very high, and rises to almost 2 000 000° K. From this one might imagine that it emits a tremendous amount of heat, but nothing could be further from the truth. Scientifically, 'temperature' is defined by the rate at which the various atoms and molecules move around; the faster the speeds, the higher the temperature. In the corona the speeds are very great, but there are so few particles that the actual 'heat' is negligible. We are still not sure why the coronal temperature is so high, but on the whole it seems likely that once again we are dealing with some sort of magnetic phenomenon.

Unfortunately, the Moon's shadow is only just long enough to touch the Earth, and so eclipses as seen from any one particular locality are relatively rare. If the orbits of the Earth and the Moon lay in the same plane, we would have a solar eclipse at every new moon, but in fact the lunar orbit is inclined to ours at an angle of just over 5 degrees, so that on most occasions the new moon passes unseen either above or below the Sun in the sky. To produce an eclipse, the new moon must be very near a node – a node being the point where the Moon's path cuts the plane of the ecliptic. Moreover, a total eclipse is a rushed affair; it can never last for as long as eight minutes, and is generally much briefer. To either side of the zone of totality there is, of course, a partial eclipse.

It so happens that the Sun, the Moon and the nodes arrive back at almost the same relative positions in a period of approximately 18 years 11 days, known as the Saros, so that any eclipse is quite likely to be followed by another eclipse 18 years 11 days later. The conditions are not identical, but the Saros is a reasonable guide, and was good enough for the ancients to make predictions. Thus Thales of Miletus was able to forecast the eclipse of 25 May 585 BC, which is said to have stopped a battle between the armies of the Lydians and Medes; the combatants were so alarmed at the sudden darkness that they concluded a hasty peace. Not that the sky is completely dark during totality; but it becomes dusky enough for planets and bright stars to be seen.

There are many phenomena associated with total eclipses – for example, Baily's Beads, bright points seen just before and just after totality when the sunlight passes through valleys on the Moon's limb. Then there are shadow bands, wavy lines on the Earth's surface also seen just before and just after totality.

All in all, there is nothing to rival the glory of a total eclipse; if you have the chance of seeing one, do not hesitate. The last completely visible total eclipse in England was that of 11 August 1999; the next will not be until 23 September 2090, though other parts of the world are more fortuntate. Thus there will be totalities on 22 July 2009, visible from the Indian Ocean and Sri Lanka areas, and on 11 July 2010, visible from the Pacific Ocean areas.

It is no surprise to find that the Sun sends out radiations at all wavelengths. Studies of solar X-rays have to be carried out from space vehicles, and these methods have been invaluable; for instance we can now track the so-called 'coronal holes', regions of exceptionally low density in the corona, which are important because they provide escape routes for the solar wind – made up of streams of low-energy particles sent out by the Sun all the time. There have been special probes, such as Ulysses, which was put into a path enabling it to study the poles of the Sun, which we can never examine properly from Earth because we see them at such an unfavourable angle. The polar regions are of special significance, because it is here that the lines of the Sun's magnetic field are open rather than closed, making it easier for the solar wind particles to break free. Predictably, the Sun is also a strong source of radio waves, and here at least a great deal of research can be carried out from ground level.

Finally, it is worth saying something about a problem which puzzled astronomers for years. Theory predicts that as the Sun radiates, it should release vast numbers of particles known as neutrinos, which are excessively difficult to detect because they have no electrical charge and virtually no mass – perhaps none at all. Deep in Homestake gold-mine in South Dakota, Ray Davies and his team have built what is certainly the world's strangest 'telescope'. It consists of a large tank of over 460 000 litres (101 200 gallons) of cleaning fluid, and it is buried in a disused mine-shaft 1.6 km (1 mile) below ground. The fluid is rich in chlorine, and it is known that if a neutrino scores a direct hit on a chlorine nucleus it can change the chlorine into a special form of another gas, argon, which is easy to identify. The underground location is essential because if the experiment were set up on or near the Earth's surface it would be wrecked by other particles, known (misleadingly) as cosmic rays, which are bombarding us from all directions all the time, and would produce the same effects. Luckily, the cosmic rays cannot penetrate down to Davies' tank of cleaning fluid.

The startling result was that far fewer neutrinos were detected than had been anticipated. For a long time this remained a mystery, but it now seems that the problem has been solved. There are three types of neutrinos, and of these only the so-called 'electron' neutrinos can be recorded at Homestake and similar installations elsewhere. What apparently happens is that during the journey from the Sun to the Earth, the electron neutrinos change into the other types which cannot be recorded, and so are missed. Subsequent experiments, with more elaborate detectors, have provided ample confirmation of this. Once again the Sun has been of immense help to physicists.

Always remember, too, that the Sun is the only star which is sufficiently close to us to be examined in great detail, so that it is the main key in our quest for knowledge of the other stars. Solar physics has become one of the most vital branches of modern astronomy, and there are many observatories which concentrate solely upon it. We have come a long way since Herschel claimed that the Sun is a cool, habitable world.

7

the Moon

In this chapter you will learn:
- about the Moon, its phases and its effects on our tides
- about the Moon's surface and how lunar craters were formed
- about the eclipses of the Moon.

Just as the Sun rules the day, so the Moon is dominant at night, at least for part of every month; full moonlight can appear so brilliant that it is not always easy to appreciate that it would take roughly half a million full moons to equal the light of the Sun. Yet the amount of heat sent to us from the Moon is negligible, so that it is quite safe to look straight at it through a telescope or binoculars. (Recently there has been a report of eye damage due to staring at the full moon through a telescope. I admit to being somewhat sceptical about this; in any case there can be no problem at all unless you deliberately stare for a very long time – and the same could be said of looking straight at a searchlight. It is, surely, a question of common sense.)

Imposing though it looks, the Moon is a junior member of the Sun's family. It is smaller than the Earth, with a diameter of only 3472 km (2158 miles), and it has only 1/81 of the Earth's mass, so that its gravitational pull is much weaker (see Figure 7.1); the escape velocity is a mere 2.4 km (1½ miles) per second, so that the Moon has been unable to hold on to any atmosphere it may once have had, and is today airless, waterless and lifeless. The dark patches still called 'seas' have never had any water in them, though it is true that they were once oceans of liquid lava.

We cannot be sure about the origin of the Moon. The old idea that it simply broke away from a rapidly spinning Earth has been given up – mathematically, it is untenable – and there are also difficulties in assuming that it was once an independent body which came too close to the Earth and was unable to break free again. There is growing support for the idea that at an early stage in the evolution of the Solar System there was a collision between the Earth and a large wandering body, hurling around a vast amount of débris which then collected to form the Moon. At least we know that the two bodies are of much the same age; the rocks brought home by the Apollo astronauts and the automatic Russian probes show that both are of the same vintage.

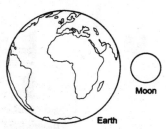

Figure 7.1 Earth and Moon compared

It is often said that 'the Moon moves around the Earth'. This is true enough, but there is one modification, because strictly speaking the Earth and Moon move together around the barycentre, which is the common centre of gravity of the system. But since the Earth is 81 times as massive as the Moon, the barycentre lies deep inside the terrestrial globe, and the simple statement is good enough to satisfy anyone except a mathematician. The lunar orbit is not circular; the distance from Earth ranges between 356 300 km (221 460 miles) at its closest (perigee) out to 406 600 km (252 700 miles) at its furthest (apogee).

The orbital period is 27.3 days. The cause of the regular phases, or apparent changes of shape, is easy to explain. When the Moon is more or less between the Earth and the Sun, its dark or night side is facing us, and we cannot see it at all – except when the alignment is exact, and we have a solar eclipse (as we have noted, this does not happen every month, because of the

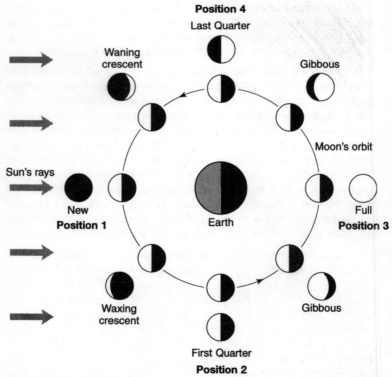

Figure 7.2 Phases of the Moon

5-degree tilt of the Moon's path). This, and not the slender crescent seen soon afterwards in the evening sky, is the true new moon. In position 2 in Figure 7.2 half the sunlit side faces us; rather confusingly this is known as First Quarter, because the Moon has then completed one-quarter of its journey. In position 3 we see the whole of the sunlit face (full moon) and in position 4, half again (Last Quarter). Between half and full the Moon is said to be gibbous. Because the Earth and the Moon are moving together around the Sun, the interval between one new moon and the next, known as the synodic period, is 29½ days rather than just over 27; Figure 7.2 should make this clear.

When the Moon is in the crescent stage, the night side can often be seen shining faintly. This is due to light reflected on to the Moon from the Earth, and is therefore known as the earthshine. Its cause has been known for a long time; Leonardo da Vinci may have been the first to explain it. Country folk call it 'the Old Moon in the Young Moon's arms'. And while on the subject of country lore, it is worth adding that so far as we can tell, the Moon has absolutely no effect on the weather. Quite apart from anything else, new moon and perigee do not often coincide; there is no reason why they should.

The Moon is a slow spinner. Our rotation period is 24 hours; that of the Moon is 27.3 days – exactly the same as its orbital period. This may sound a strange coincidence, but there is a straightforward explanation for it. Initially the Earth and the Moon were viscous, and raised tides in each other. The tides raised by the Earth on the Moon were much the more powerful, because of the Earth's greater mass, and the effect was to slow down the Moon's rotation as it span; the Earth tried to keep a 'bulge' pointing in our direction. Eventually the rotation had stopped altogether relative to the Earth – but not relative to the Sun, so that on the Moon's surface day and night conditions are the same everywhere. A lunar day is almost as long as two Earth weeks.

It follows that the same side of the Moon is always turned towards us, and that there is part of the surface which we can never see – a fact which used to infuriate pre-Space Age observers (such as myself). However, things are not quite so bad as they might seem. Though the Moon turns on its axis at a constant speed, it does not move around us at a constant speed; following Kepler's Laws, it moves quickest when closest-in. Each month, therefore, the position in orbit and the amount of spin become out of step, and we can peer a little way around alternate mean limbs. This, together with other less important effects of the same kind, known as librations (see Figure 7.3),

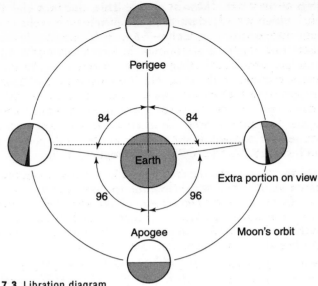

Figure 7.3 Libration diagram

means that altogether we can examine a grand total of 59 per cent of the whole of the surface; it is only 41 per cent which is permanently hidden. With the circumlunar flight of the unmanned Russian spacecraft Lunik 3, in 1959, we had our first direct view of the Moon's far side. Predictably, it proved to be essentially similar to the areas which we have always known, though there were differences in detail.

The main effect of the Moon upon the Earth concerns the tides. As a start, let us imagine that the whole Earth is covered with a shallow uniform ocean, and that both the Earth and the Moon are standing still. In Figure 7.4, which is not to scale, we will have a high tide at point A and another high tide at point B, with low tides at C and D; the Moon's gravity is heaping the water up at A, where the pull is strongest. Of course the solid land is being pulled as well, but land is much more difficult to shift than water, so that the heaping-up is less.

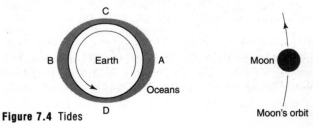

Figure 7.4 Tides

This is all very well, but why should there also be a high tide at B? It is slightly misleading to say simply that the solid globe is being 'pulled away' from the water, so for the moment let us assume that we have a situation in which the Earth and the Moon are absolutely alone in the universe, and are falling towards each other. Point A, which is closest to the Moon, will be subject to a greater accelerating force than the average, and so the water will be bunched upward. At B, the reverse will happen. The acceleration will be less than average, so that the water in that region will tend to be 'left behind', bulging away from the Moon and producing a similar high tide.

Next, assume that the Earth and the Moon are keeping the same distance apart, but that the Earth is spinning around once in 24 hours. Obviously the water-heaps will not spin with it; they will try to keep 'under' the Moon. Therefore, each bulge will sweep around the Earth once a day, so that every region will have two high tides and two low tides in each 24 hours.

However, there are many complications. First, the Earth is not covered with a uniform water-shell, and the outlines of the seas and continents play a major role. Second, the Moon is moving along in its orbit, so that the water-heaps shift slowly as they follow the Moon around. Also, the waters take time to heap up, so that the maximum tide does not fall directly below the Moon; there is an appreciable lag. All in all, tidal phenomena are surprisingly complicated.

Neither can we neglect the Sun, which also raises tides. The tides are much weaker than those of the Moon, because the Sun is 400 times as remote, and the main tide-raising force is due to the difference in pull on opposite sides of the Earth. When the Sun and Moon are pulling in the same sense, at new or full moon, the tides are strongest (spring tides; the term has nothing to do with the season of spring), while when the Sun and Moon are pulling against each other, at half-moon, we have the weaker neap tides. Anyone who lives on the coast, as I do, will be very conscious of the difference in strength.

I appreciate that this account of tidal phenomena is hopelessly over-simplified, but to go into real detail would take many pages, so let us now turn to the lunar world itself.

Even a casual glance at the Moon will show the light and dark areas; who has not heard of the Man in the Moon? It is said that the patterns do give a vague impression of a human form, though I admit that I have never myself been able to make it out.

Originally it was thought that the bright areas were land and the dark areas water; we still use romantic names such as the Sea of Tranquillity, the Bay of Dews and the Ocean of Storms, though at an early stage in the telescopic era it was realized that the so-called seas are bone-dry. The Latin names are always used; thus the Sea of Tranquillity becomes the Mare Tranquillitatis, the Bay of Dews is Sinus Roris, and so on.

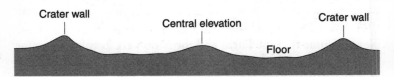

Figure 7.5 Profile of lunar crater

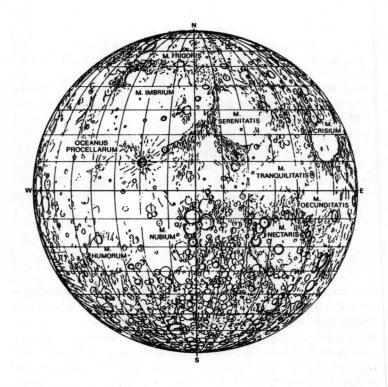

Figure 7.6 Outline lunar map

Some of the seas or *maria* are fairly regular in form; the largest of these, the Mare Imbrium (Sea of Showers) is bounded in part by high mountain ranges, the Lunar Apennines and Alps. Ranges are often given terrestrial names, and some of the peaks are lofty, rising to well over 6000 m (20 000 ft); there are also many isolated hills and mountains. Level country is rare on the Moon.

Of course the whole scene is dominated by the craters (see Plate 2), which range from huge enclosures well over 240 km (150 miles) across down to tiny pits at the limit of visibility as seen from Earth. Many of the craters have terraced walls, together with central mountains or groups of mountains; thus the superb Theophilus, at the edge of the Mare Nectaris (Sea of Nectar) is 103 km (64 miles) across and has a massive central mountain complex. The highest peaks on the wall-crest rise to 5486 m (18 000 ft) above the deepest part of the sunken floor, but to a much lower height above the outer surface, and in fact a lunar crater seen in profile would resemble a shallow saucer much more than a steep-sided well (see Figure 7.5). Neither does a central peak ever attain the height of the outer rampart, and it is probably apt to call a lunar crater a sunken plain, though there is one formation, the 87-km (54-mile) Wargentin, which is lava-filled and takes the form of a plateau.

The craters were named by the Italian astronomer Giovanni Riccioli, who drew a lunar map in 1651 (see Figure 7.6). His method was to christen craters after famous persons, usually (though not always) astronomers. The system has been followed through to the present day, and has been extended, but naturally Riccioli 'used up' all the main craters, so that later astronomers such as Sir Isaac Newton and Sir William Herschel have had to make do with second best. Galileo Galilei is represented by a very obscure crater in the Oceanus Procellarum, but this was deliberate; Riccioli had no faith in the Copernican system, and, in his own words, 'flung Galileo into the Ocean of Storms'. However, Riccioli did relent sufficiently to give Copernicus one of the most splendid formations on the Moon. It lies in the Mare Nubium (Sea of Clouds); it is well-formed, 90 km (56 miles) across with superbly terraced walls, and near full moon it is seen to be the focal point of a system of bright streaks or rays which stretch out for long distances in all directions. There is an even more extensive ray-centre in the southern uplands of the Moon, issuing from the crater named in honour of Tycho Brahe. The rays are surface deposits; they cast no shadows, and are seen well only when the Sun is high above them.

Not all craters are similar to Tycho and Copernicus. For example Plato, which is 97 km (60 miles) across and lies in the uplands between the Mare Frigoris (Sea of Cold) and the Mare Imbrium, has an iron-grey floor which is almost devoid of detail. Though it is circular, it is well away from the apparent centre of the Moon's disk, and is drawn out into the shape of an ellipse. All craters near the limb are foreshortened, and in the libration regions – that is to say, the areas which are periodically tipped in and out of view – it is often difficult to distinguish a crater-wall from a ridge. Before the coming of the space probes, maps of these regions were of low accuracy.

Minor features of the Moon include domes, which are low swellings often crowned by summit pits; ridges, some of which snake across the maria; valleys, and crack-like features known as rills, rilles or clefts, some of which are prominent enough to be seen with very small telescopes. Many of them are in part crater-chains, though others are fairly obviously collapse features, and look superficially like cracks in dried mud.

The distribution of the craters is not random; when one formation breaks into another it is almost always the smaller crater which intrudes into the larger. There are pairs of craters, groups, and chains of vast formations which are clearly of different ages. Some craters have been so broken and distorted that they are barely recognizable, and this applies also to some of the major walled plains; for example Bailly, in the far south, is more than 290 km (180 miles) in diameter, but has been fittingly described as 'a field of ruins'.

How were the craters formed? For many years there were arguments between those observers who regarded them as volcanic, and those who believed them to be due to meteoritic impact. We now know that the impact theory is correct. (There were, of course, other theories too, such as that proposed by the Spanish engineer Sixto Ocampo. To him, the craters were the result of a nuclear war between two races of Moon-men. The fact that some craters have central peaks while others do not shows, naturally, that the two sides used different kinds of bombs!)

From the modern picture, the Moon was formed at the same time as the Earth, 4600 million years ago. The globe began to cool, and a crust was formed. Then came the 'Great Bombardment', when material rained down upon the lunar surface to produce the oldest basins, such as that of the Mare Tranquillitatis. As the Great Bombardment eased there was widespread vulcanism, with magma pouring out from below the crust and flooding the basins; craters with dark floors, such as Plato, were also flooded. The lava-flows ended rather suddenly by cosmical standards, and since then there has been no major activity, apart from the formation of occasional impact craters such as Tycho. The Moon's rotation has been synchronous since a very early stage and the topography of the far side differs in some ways from that of the Earth-turned hemisphere; in particular there are no 'seas' of the size of the Mare Imbrium.

Mild signs of activity are sometimes seen, indicating the release of gas from below the crust; these 'Transient Lunar Phenomena' are elusive, and all major activity on the Moon belongs to the remote past.

I have already said something about the space missions, which began with the automatic Russian Luniks of 1959 and were followed by the US Surveyors and Orbiters, further Russian unmanned missions, and then of course the Apollo landings (see Plate 6), culminating in the touch-down of the Eagle, from Apollo 11, in July 1969. I doubt whether anyone has bettered Buzz Aldrin's description of the lunar scene as 'magnificent desolation'. The sky is black even in the daytime; there is no 'weather', and no sound. The flags set up by the astronauts do not flutter; how could they, in the absence of wind? Various artifacts left on the surface will not deteriorate; there are various crashed space-craft, the bottom stages of Apollo modules, three 'Moon cars' in which the astronauts of Apollos 15, 16 and 17 drove themselves around, and of course the scientific recording instruments, now (sadly) switched off for reasons of economy. No doubt all these will be collected in due course and taken off to lunar museums ... and there is nothing far-fetched about a Lunar Base. If we wanted to do so, we could probably build it within the next couple of decades, and it would be of immense value, not least as a medical research centre.

In 1994 a small photographic probe, Clementine – named after the character in the old mining song who is 'lost and gone forever' – proved to be very successful. In particular, Clementine surveyed the lunar poles, and it has been suggested that there is ice inside some of the craters whose floors are permanently in shadow, so that they remain bitterly cold. From the outset I had the gravest reservations about this; in particular, what could be the origin of the ice? A crashing comet would generate far too much heat. Moreover, none of the rocks brought back by the astronauts contained any signs of hydrated material.

In January 1998 another probe, Prospector, was launched and put into lunar orbit, mainly to search for ice. The results were negative and eventually, on 31 July 1999, Prospector was deliberately crashed inside a crater where ice, if it existed at all, would be present. The débris was again devoid of any icy material and, all in all, it seems that the whole idea of ice inside polar craters must be abandoned. In 2007 came the first unmanned moon missions by Japan (Kaguya) and China (Chang-1); we may expect many more probes in the near future as we enter a new era of lunar exploration.

Telescopically, the lunar aspect changes with bewildering rapidity, not because it actually does so but because of the changing angle of sunlight. When a crater is near the terminator, or boundary between the daylit and night hemispheres, it is wholly or partly shadow-filled, and may be very prominent; near full moon the shadows virtually vanish, and the scene is dominated by the bright ray systems, mainly those of Tycho and Copernicus, so that even large walled plains may be hard to identify unless they are exceptionally bright or exceptionally dark. Plato is always recognizable because of the greyness of its floor, and so are Grimaldi and Riccioli near the western limb, while the 37-km (23-mile) crater Aristarchus, in the Oceanus Procellarum, is so brilliant that unwary observers have mistaken it for a volcano in eruption. It does not take long to learn your way around the lunar surface, but remember that the best views are obtained during the crescent, half and gibbous phases; full moon is the very worst time for the beginner to start observing.

When the Moon is full, it may sometimes pass into the shadow cast by the Earth, and the result is a lunar eclipse (see Figure 7.7); this does not happen every month, again because of the 5-degree tilt of the Moon's orbit. Lunar eclipses may be either total or partial, and are more common than solar eclipses as seen from any particular site – because an eclipse of the Moon can be seen from any place where the Moon is above the horizon, whereas to see an eclipse of the Sun you have to be in just the right place at just the right time. It cannot be said that a lunar eclipse is spectacular, but it is worth watching; the supply of direct sunlight is cut off, and the Moon becomes dim and often coppery until it passes out of the shadow again. In general it does not disappear completely, because some of the Sun's rays are bent on to it by way of the shell of atmosphere surrounding the Earth, but everything depends on conditions in the upper air, through which all the sunlight reaching the eclipsed Moon has to pass. When the upper regions are laden with dust and ash – as for example in the months following the eruption of Mount Pinotubo in the Philippines – the eclipse may be so 'dark' that the Moon is difficult to find at all; at other eclipses there are lovely colours, and there is great scope for photographers, particularly as a lunar eclipse is a comparatively leisurely affair. Totality may last for well over an hour.

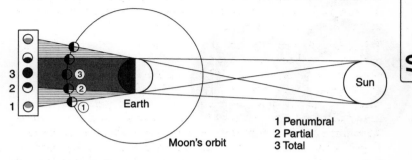

Figure 7.7 Theory of lunar eclipse

There will be total lunar eclipses on 21 December 2010, 15 June and 10 December 2011, and 15 April and 8 October 2014.

There are also times when the Moon passes in front of a star, and hides or occults it. Because of the lack of lunar atmosphere, the star shines steadily until the instant when it is covered up by the advancing limb, when it snaps out as suddenly as a candle-flame in the wind. Planets can also be occulted now and then, though here the immersion and emersion are gradual, because a planet appears as a disk rather than a point source of light.

After Apollo, there seemed to be a general feeling that we had learned all that we wanted to know about the Moon, and that there was little point in studying it further. This is quite wrong. There are still many problems to be cleared up, and in any case the Moon has lost none of its magic; it is our faithful companion in space, and our skies would indeed be dull without it.

08

the Sun's family

In this chapter you will learn:
- about the members of the
  Solar System
- how the members of the
  Solar System move and
  how they differ from each
  other.

There was a time, not so very long ago, when it was generally thought that the Solar System was unique, or at least so exceptional that the chances of finding 'other Earths' in the Galaxy was minimal. Nowadays we think differently, because our theories have changed.

Much depends upon how the planets were formed. Initially there was great support for the so-called Nebular Hypothesis, proposed by the great French scientist Pierre-Simon Laplace in the 1790s. He assumed that the system began as a rotating gas-cloud, which shrank under the influence of gravitation; regular rings were thrown off, each one of which condensed into a planet, while the central part of the cloud became the Sun. However, there were so many mathematical objections that the idea was abandoned and was replaced by a crop of 'tidal theories', of which the most celebrated was championed by Sir James Jeans, a well-known popularizer of astronomy during the inter-war years (he was also a leading astrophysicist, and only in his later career did he change over to writing books for beginners). According to Jeans, the planets were pulled off the

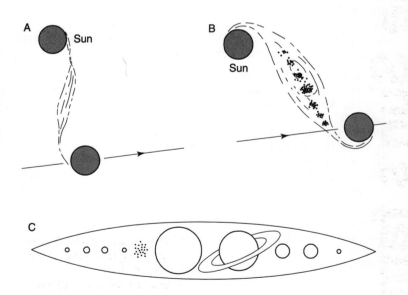

Figure 8.1 Jeans' tidal theory. Star passes close to sun, pulling material from it (A and B). Planets form from this material (C)

Sun by the action of a passing star; a cigar-shaped tongue of material was produced, and this broke up to form the planets. Predictably, the largest planets, Jupiter and Saturn, are in the middle of the system, where the thickest part of the cigar would have been (see Figure 8.1).

If this had been valid, then Solar Systems would have been vanishingly rare, because the stars are so widely spaced that close approaches seldom occur, and it would have been quite likely that our planetary system would have been the only one in the Galaxy. But again the mathematicians swarmed to the attack, and eventually the tidal theory joined Laplace's gaseous rings on the scientific scrap-heap.

We now have a much more plausible theory, which, ironically, is closer to Laplace's than to Jeans'. The planets are assumed to have condensed out of a cloud of material associated with the youthful Sun, so that they are of similar age. This explains why the Solar System is divided so sharply into two parts. Close to the Sun the temperatures were high, so that light gases such as hydrogen and helium were driven away, and the inner planets were left with only the heavier elements. Further out the temperatures were lower, so that the lightest elements could be retained – which is why the giants, unlike the inner members of the system, are not solid and rocky. True, a giant planet may have a solid core, but most of their globes are liquid, with deep atmospheres rich in hydrogen and helium.

Though the planets shine only by borrowed sunlight, their reflecting powers or albedoes are very different. Thus cloud-covered Venus reflects about 76 per cent of the sunlight falling on it, while the albedo of rocky Mercury is a mere 6 per cent (much the same as the average albedo of the Moon). And because the planets are of different sizes, and lie at different distances from us, they show a wide range in brilliancy. The first five must have been known from the dawn of human history; Venus and Jupiter are always much brighter than any of the stars, and so is Mars at its best, while Saturn is prominent enough, and even Mercury can be quite conspicuous. The ancients knew Mercury well, even though initially they believed that the 'evening Mercury' and the 'morning Mercury' were two different bodies – an error also made with regard to Venus.

A table summarizing the main planetary data is given in Appendix 1, but perhaps a scale model will be useful here, so let us represent the Sun by a football and put it in the middle of a large field (see Figure 8.2). Mercury will then be a grain of sand

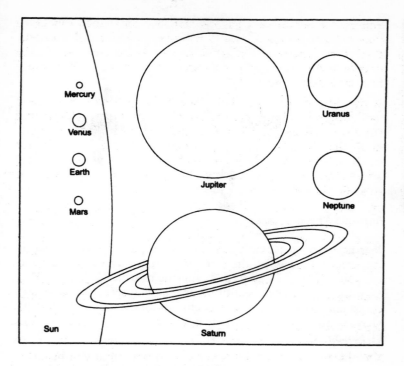

Figure 8.2 Sizes of the planets compared with the Sun

at approximately 18 m (60 ft) from our model Sun; Venus, a pea at 37 m (120 ft); the Earth, another pea at 50 m (165 ft); Mars, a rice-grain at 76 m (250 ft); Jupiter, a golf-ball at 259 m (850 ft); Saturn, a table-tennis ball at 472 m (1550 ft); Uranus, a grape at 975 m (3200 ft), and Neptune another grape at 2255 m (7400 ft). The nearest star will be a cricket ball at over 12,870 km (8000 miles), so that if we want to include it in our model we will have to choose a very large field indeed!

Next, let us discuss the ways in which the planets seem to move. Mercury and Venus have their own way of behaving, and show lunar-type phases from new to full. In Figure 8.3 I have shown the orbit of Venus, and I have assumed that the Earth is stationary. In position 1 Venus is at inferior conjunction, i.e. between the Earth and the Sun, so that its dark hemisphere is turned towards us, and normally we cannot see it. If the lining-up is exact, Venus appears in transit as a black disk crossing the Sun's face, but because of the 3.4-degree tilt of Venus' orbit this

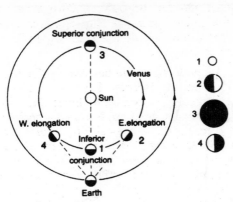

Figure 8.3 Phases of Venus

does not happen often. Transits occur in pairs at an interval of eight years, after which there are no more for over a century; the last pair fell in 1874 and 1882, and the next in 2004 and 2012. I saw the 2004 transit beautifully from my observatory in Selsey; Venus showed up as a prominent, jet black disk. Transits of Mercury are more frequent; the next will be on 9 May 2016 and 11 November 2019.

Now back to Figure 8.3. As Venus moves along it will begin to show up as a crescent in the morning sky, and by the time it reaches position 2 it will be at greatest elongation, appearing as a half-moon and often rising above the horizon several hours before the Sun. Yet it is further away from us than when it was new, and the apparent diameter is smaller. Oddly enough, Venus is at its brightest when in the crescent stage, and the phase can then be seen easily with binoculars; exceptionally keen-sighted people can make it out with the naked eye.

At position 3 Venus is full, but as it is then to all intents and purposes behind the Sun it is out of view (superior conjunction). The phases are then repeated in the reverse order: gibbous, half (position 4) and then crescent before returning to new. Curiously, the moment of exact half-phase, or dichotomy, does not usually coincide with the time of elongation. When Venus is waxing, dichotomy is a day or two late; when Venus is waning in the evening sky, dichotomy is slightly early. This was first noticed by Johann Schröter in the 1790s, and I once christened it the 'Schröter effect', a term which has come into general use. Of course it does not mean that Venus is out of position, or that there is anything wrong with the calculations; it is due to effects of Venus' dense, cloudy atmosphere. With Mercury, where the maximum elongation from the Sun is never as great as 30 degrees, there is no Schröter effect, because Mercury has practically no atmosphere.

The superior planets, which lie beyond the Earth's orbit in the Solar System, behave differently. Figure 8.4 shows the path of Mars, where the orbital period is 687 Earth days. When the Sun, the Earth and Mars are lined up, Mars is opposite to the Sun in the sky, and is said to be at opposition, so that it is best placed for study. A year later the Earth has completed one circuit, but Mars, moving more slowly in a larger orbit, has not had time to do so, and the next opposition is delayed until the Earth has 'caught Mars up'. The synodic period of Mars is therefore longer than a year; on average it is 780 days, so that there will be oppositions in 2010 and 2012 but not in 2011 and 2013. And because the Martian orbit is less circular than ours, not all oppositions are equally close. At times, as in 1988, the distance between the Earth and Mars may be reduced to around 58 million km (36 million miles); at other oppositions, such as those of 1995 and 1997, the minimum distance is greater than 97 million km (60 million miles). The next really close opposition was in 2003, when for a brief period Mars outshone even Jupiter.

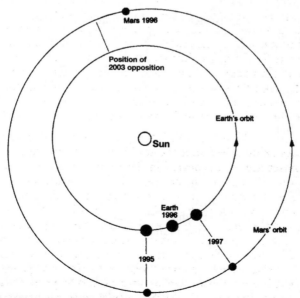

Figure 8.4 Oppositions of Mars

A superior planet cannot pass through inferior conjunction, but when on the far side of the Sun it does reach superior conjunction.

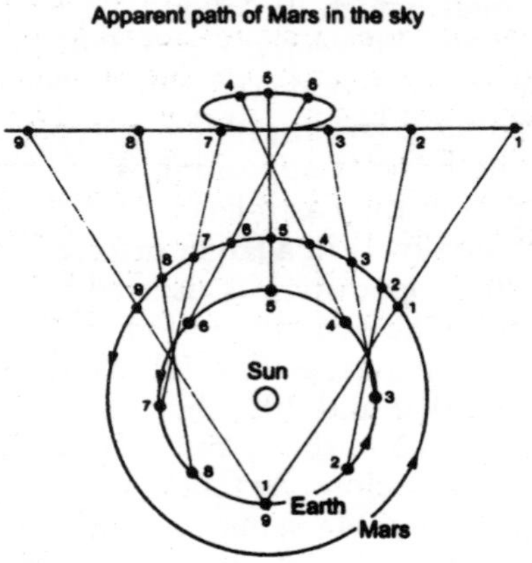

Apparent path of Mars in the sky

Figure 8.5 Retrograde motion of Mars

At right angles to the Sun it is said to be at quadrature. Mars may then show a distinct phase, like that of the Moon a day or two from full, but the phases of the other planets are inappreciable. Moreover, because the outer planets are so far away, the Earth takes less time to 'catch them up', and their synodic periods are shorter; indeed, oppositions of Neptune occur only a couple of days later each year.

A superior planet generally moves from west to east against the stars, but not always, as Figure 8.5 shows; I have again restricted it to Mars, but the principle applies equally to Jupiter and the rest. Some time before opposition, when the Earth is starting to 'pass' Mars, there is a change; Mars reaches a stationary point, and then starts to move in a retrograde or east to west direction against its background. Then there is another stationary point, after which the planet resumes its eastward march. In fact, a superior planet seems to describe a long, slow 'loop' against the stars.

All the planets, and all the asteroids, move around the Sun in the same sense as the Earth. This is only to be expected, but, more surprisingly, they do not all rotate in the same sense (see Figure

8.6). Venus is 'upside-down' from our point of view; it spins from east to west, so that if you could stand on the surface of the planet and see the Sun you would find that sunrise would be to the west and sunset to the east (though this is purely academic, since an observer on Venus would be unable to see anything above apart from a cloud-laden sky). Uranus is odd, because its rotational axis is tilted to the perpendicular by more than a right angle, leading to a very peculiar calendar. The reasons for this are unknown. The idea that a planet of this size could be struck by a massive body and literally tipped over does not sound in the least plausible and it is more likely that inclination was caused by mutual interactions between the giant planets in the early history of the Solar System.

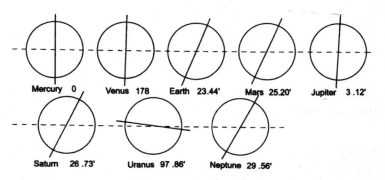

Figure 8.6 Axial tilts of the planets

The escape velocity of a planet depends upon its mass; thus Jupiter has an escape velocity of 60 km (37 miles) per second, while that of Mercury is only 4 km (2.6 miles) per second, not a great deal higher than that of the Moon. The ability to retain an atmosphere is linked with the escape velocity, because too feeble a pull means that any air will leak away – as has happened in the case of the Moon. It is no shock to find that Mars has a painfully thin atmosphere and Mercury almost none. Titan, the largest satellite of Saturn, has an atmosphere which is denser than that of the Earth even though its escape velocity is about the same as that of the Moon; the very low temperature of Titan is responsible. Atoms and molelcules move slowly under cold conditions; if Titan became much warmer, its atmosphere would escape.

There is one other fact which at first sight sounds surprising. The surface gravity of a planet depends not only upon its mass but also upon its diameter, because a body behaves as though all its mass is concentrated at its centre – and the further away you are from the centre, the weaker the pull. This is why Mercury and Mars have the same surface gravity; Mars is the more massive of the two, but it is also larger, so that the surface of the planet is further away from its centre. If you could stand upon Saturn, you would find that you would 'weigh' little more than you do on Earth.

Most of the planets are attended by satellites, but the various families are very unlike each other. Venus and Mercury are solitary travellers in space. Mars has two tiny moons, Phobos and Deimos, which are mere lumps of rocky material and are certainly ex-asteroids captured by Mars in the remote past. Jupiter has four satellites of planetary size, plus a host of midgets. Saturn has one large satellite (Titan), several of medium size and others which are very small. Uranus has four medium-sized attendants and others which are much smaller. Neptune's single large satellite, Triton, is almost certainly a captured body, and all Neptune's other attendants are insignificant.

Conventionally, the smaller planets are referred to as 'terrestrial' and the others as 'giants'. Yet what about the small fry – asteroids and so on, including the swarm of asteroid-sized bodies making up the 'Kuiper Belt', beyond the orbit of Neptune? It has been decreed that these will be called Small Solar System Bodies (SSSBs) apart from Ceres, the largest of the main-belt asteroids, and Eris and Pluto, the largest of the Kuiper Belt objects, which become 'dwarf planets'. This does not seem particulary logical, but it must suffice for the moment.

In Chapter 11 I will have more to say about the less important members of the Solar System, such as asteroids and comets. Meanwhile, it is reassuring to know that the System is a stable place. The Earth's orbit is completely regular, and there is no reason to suppose that there will be any marked change until the far-distant time when the Sun uses up its store of available hydrogen fuel and swells out, with disastrous results. Of course, it is always possible that we will be hit by a wandering comet or asteroid, as Jupiter actually was in July 1994; there must have been major impacts in the past, and there is a popular theory that one strike, some 65 million years ago, caused a change in the world's climate marked enough to eliminate the dinosaurs.

However, the chances of any similar event in the near future are very slight indeed.

I am often asked why the Earth, and only the Earth, is suitable for life of our kind. The main reason is that the Earth has the right size and mass, and moves at the right distance from the Sun. As one famous astronomer commented years ago, if conditions were not suitable for us, we would not be here. We would be somewhere else!

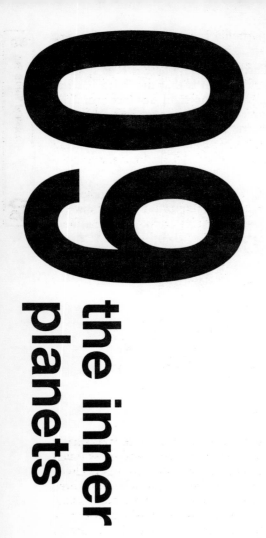

09

the inner planets

In this chapter you will learn:

- about the three planets which, together with the Earth, make up the inner part of the Solar System
- that Mercury is not too unlike the Moon
- that Venus is intolerably hot
- that Mars has some points of similarity with the Earth.

It is natural for us to be particularly interested in the planets which move in the inner part of the Solar System, because they are our nearest neighbours and are well within the range of modern rockets. Yet they are not at all like the Earth, and neither are they particularly welcoming. Even Mars, so long regarded as a possible abode of life, is not a place where we could survive under natural conditions.

In any description of the Sun's family it is conventional to begin with Mercury and then work outward. This is logical enough – so let us be conventional.

Mercury is a small world, only 4875 km (3030 miles) in diameter, and since it never comes much within 80 million km (50 million miles) of us it is never glaringly conspicuous; there are many people – even some astronomers! – who have never seen it at all (see Figure 9.1). Actually it can become surprisingly

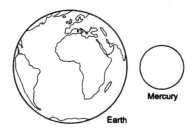

Figure 9.1 Mercury and Earth compared

bright, and can outshine any star, but it is never seen against a dark background, and no Earth-based telescope will show much in the way of surface detail. Two space-craft have made close-range studies of it. The first was Mariner 10, which made three active passes in 1974–5 before contact was lost. Next came Messenger, launched in 2004; it made its first fly-by of Mercury in January 2008, and is due to begin orbiting the planet in March 2011. Before the Mariner 10 pass, it was tacitly assumed that the surface would be similar to that of the Moon, but we had very little reliable information to guide us.

Of course the orbit was well known, with its 'year' of 88 Earth-days, but the rotation period was not. For a long time it was believed that the axial rotation period was the same as the orbital period, in which case Mercury would keep the same face turned towards the Sun – just as the Moon always keeps the same face turned Earthward, and for much the same reason.

However, Mercury's orbit is appreciably eccentric, and the distance from the Sun ranges between 69 million km (43 million miles) and only 47 million km (29 million miles), so that the orbital speed varies too, and there would have been important libration effects. There would have been a region of permanent day, an area of eternal night, and a narrow intermediate 'twilight zone', where the Sun would bob up and down over the horizon. Science fiction writers made great play of the Mercurian twilight zone, but in the 1960s it was found that the true situation was not like this at all. Infra-red techniques proved that the dark side is not nearly so cold as it would be if it never received any sunlight. Radar methods were then called in, and established that the real rotation period is 58.6 days, or two-thirds of a Mercurian year, so that all areas of the surface are in sunlight at some time or other – apart from the floors of some deep craters near the planet's poles, which are never shadow-free.

The first serious attempts at mapping Mercury were made over a century ago, by G. V. Schiaparelli in Italy, but before the flight

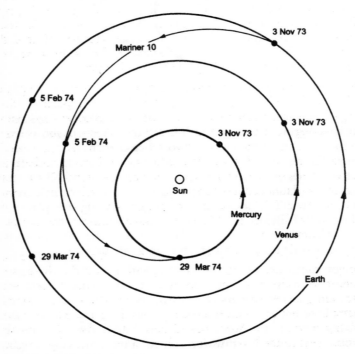

Figure 9.2 Path of Mariner 10

of Mariner 10 the best chart was due to Eugenios Antoniadi, a Greek-born astronomer based in France, who used the 84-cm (33-inch) refractor at the Meudon Observatory, outside Paris, and produced a map showing bright patches and dark areas. He even gave names to the features he drew – for example, Solitudo Hermae Trismegisti, the Wilderness of Hermes the Thrice Greatest. Alas, it was later found that the map bears very little relation to the truth, and an entirely new nomenclature has had to be introduced. This was not Antoniadi's fault; he was a brilliant observer, and he followed the sensible course of observing Mercury during daylight, when it was high above the horizon. His failure was due solely to the fact that the planet is so small and so far away that it is an excessively difficult object to see in any detail. Antoniadi was also misled by the fact that whenever Mercury is best placed for observation from Earth, the same side is presented to us.

Mariner 10 was the first probe to make use of the gravity-assist technique (see Figure 9.2). It was launched in November 1973, by-passed Venus in the following February, and made its first rendezvous with Mercury in March 1974. A second encounter took place in September 1974 and a third in March 1975, by which time the on-board equipment was starting to fail, and contact was lost on 24 March. No doubt Mariner 10 is still orbiting the Sun, and still making periodical close approaches to Mercury, but we have no hope of finding it again.

The next probe, Messenger (MErcury Surface, Space ENvironment, GEochemistry and Ranging – a typical NASA acronym!) was launched from Canaveral on 3 August 2004, and after making one fly-by of Earth and two of Venus, made its first pass of Mercury in January 2008; two more passes are needed before the probe will have slowed down sufficiently to be put into orbit around the planet, scheduled for 18 March 2011. But even in its first encounter Messenger showed itself to be a great success. It sent back images of 25 per cent of the area not surveyed by Mariner, so that we now have good pictures of over half the total surface.

Superficially the surface does look very like that of the Moon, though there are important differences in detail; there are mountains, valleys, craters, ray-systems and ridges. There are also long, winding 'lobate scarps' and what are termed 'intercrater plains', unlike anything found on the Moon. One striking feature is a huge, ringed basin, 1545 km (960 miles) across, surrounded by plains which look unusually bright, possibly because the huge impact that created the basin excavated deep material. The structure has been named the Caloris Basin, because an observer there would see the Sun directly overhead when Mercury is at perihelion, and the temperature would be around 430° C – as hot as any place on Mercury (hence the name; 'caloris' means 'heat'). Antipodal to it are hilly, grooved areas nicknamed 'WeirdTerrain'; evidently seismic waves from the Caloris impact travelled right across the globe. Crater-counts indicate that the basin was formed 3.8 to 3.9 thousand million years ago.

It seems that Mercury has an iron-rich core larger than the whole globe of the Moon, and there is a pronounced magnetic field, in which respect Mercury differs from both Venus and Mars. The crust is probably up to 30 miles thick. There is abundant evidence of violent past vulcanism, but there can be no volcanic activity now. Some craters in the polar regions, such as the 170-km (104-mile) Chao Meng-fu (latitude 87 degrees S., longitude 132.4 degrees W) are so deep that parts of their floors are always in shadow, and remain intensely cold, around −170° C. It has been suggested that water ice may have collected here, but this seems highly improbable, to put it mildly.

Mariner and Messenger confirmed that Mercury is virtually airless, which is not surprising in view of the low escape velocity and high temperature. There is a trace of atmosphere, but this is in the form of collisionless gas, and the ground density is about 1/10 000 millionth of a millibar. Any form of life there seems to be out of the question. New robot probes are being planned, but manned travel there belongs to the far future.

Venus (see Plate 5), next in order of distance, is as unlike Mercury as it could possibly be. In size and mass it is almost the equal of the Earth, and its escape velocity is only slightly lower than ours, so that logically it might be expected to have the same kind of atmosphere – but this is emphatically not so. Well before the Space Age it was known that the atmosphere of Venus is so dense and so cloud-laden that it hides the surface permanently. There is no such thing as a sunny day on Venus!

Not many decades ago, Venus was generally termed 'the planet of mystery', even though it is our nearest neighbour apart from the Moon and occasional comets and asteroids. We did not even know the length of the axial rotation period, and it was suggested that the rotation might be synchronous, i.e. the same as the orbital period of 224.7 Earth days. If so, then there would have been one permanently lit hemisphere and one hemisphere which would be permanently dark; neither would there be marked libration effects, because the orbit of Venus is practically circular.

All that could really be done was to analyse the upper atmosphere, using spectroscopic methods. It was found that the main constituent is carbon dioxide, and since this is a heavy gas which would be expected to sink it was reasonable to assume that carbon dioxide made up most of the atmosphere down to ground level. Since carbon dioxide acts in the manner of a greenhouse, blocking in the Sun's heat, it followed that Venus was likely to be a very torrid sort of world.

Yet opinions differed, and there was also a theory according to which the clouds contained a great deal of H_2O. It was even claimed that the surface might be largely ocean-covered, in which case the atmospheric carbon dioxide would have fouled the water and produced seas of soda-water. Another intriguing theory made Venus very similar to the Earth of over 200 million years ago, when the coal deposits were being laid down; there would be swamps, luxuriant vegetation of the fern and horsetail variety, and primitive life-forms such as giant dragonflies. If so, then Venus might presumably evolve in the same way that the Earth has done.

Little more could be learned from studies of the upper clouds, and the shadings seen from time to time were too vague to give any real clue as to the length of the rotation period – though French observers maintained that the rotation was equal to about four Earth days, in a retrograde or east-to-west direction. The faint luminosity of the night side during the crescent stage, seen by most regular observers of the planet and christened the Ashen Light, was another problem. Few people agreed with the nineteenth-century German astronomer Franz von Paula Gruithuisen that it could be due to illuminations lit by the local inhabitants to celebrate the election of a new Emperor, but nobody could quite explain it.

Then, in 1962, the American probe Mariner 2 by-passed Venus at less than 35 400 km (22 000 miles), and gave us our first reliable information. The surface proved to be very hot indeed; we now know that the maximum temperature is almost 500° C, above even that of Mercury. The atmosphere really is almost pure carbon dioxide, at a ground pressure of roughly 90 times that of the Earth's air at sea-level; there is no detectable magnetic field, and those shining clouds are rich in sulphuric acid. The rotation period is 243 days, longer than Venus' 'year', and is retrograde. Yet the French had been right, too; the upper clouds race around in four days, so that we have a classic case of what is termed super-rotation. All ideas of a pleasant, oceanic Venus had to be abandoned.

The Russians were next in the field, and their automatic landers were able to send back pictures directly from the surface – initially with Venera 7, in December 1970, which survived for almost half an hour before being silenced (earlier Veneras had been literally squashed during their descent through the thick lower atmosphere of the planet). The scene could hardly have been more hostile: a rocky, scorching landscape, about as bright as 'Moscow at noon on a cloudy winter day', to quote the official Soviet source. The rocks are orange, because they reflect the light from the clouds, and there is little wind on the surface, though higher in the atmosphere there must be sulphuric acid rain which evaporates before it can reach the ground (See Plate 7).

However, the main information has come from space-craft which have been put into closed orbits around Venus and studied the surface by radar. Magellan which was launched in 1989, continued to send back data until 1994.

Venus is a world of plains, highlands and lowlands; a vast rolling plain covers 60 per cent of the surface, with highlands accounting for only 8 per cent. There are two large upland areas, Ishtar Terra in the northern hemisphere and Aphrodite Terra mainly in the south; Aphrodite is about the size of Africa, while the area of Ishtar is much the same as that of Australia. There are also some smaller highland areas, notably Beta Regio, where we find a massive shield volcano – Rhea Mons – which may quite well be erupting at the present time; another probably active area, adjoining Aphrodite, has been nicknamed the Scorpion's Tail.

In one way the situation on Venus differs markedly from that on Earth. The terrestrial crust is divided up into various large plates which drift around over the hot mantle, so that a volcano appearing at a 'hot spot' near the boundary of a plate will stay there indefinitely. Thus one of the main Hawaiian volcanoes, Mauna Kea, has been carried away from the underlying hot spot, and has become extinct; the original site is now occupied by another shield volcano, Mauna Loa, which is very active indeed. On Venus there are no comparable plates, and no systematic movement of the crust, so that when a volcano is born it can stay above its hot spot for a very long time, giving it the chance to become truly enormous.

The radar maps show features of all kinds. Craters abound, some of them very large indeed – well over 240 km (150 miles) in diameter. There are ridges, valleys, channels and strange volcanic structures which look superficially rather like mushrooms. There are the curious 'arachnoids', circular volcanic features surrounded by complex troughs, and there are circular lowland areas which are characterized by intersecting ridges and grooves. These used to be call parquet areas, but the term was regarded as unscientific, and the regions are now known officially as tesserae. The deepest valley, Diana Chasma in the Aphrodite area, goes down for over 1.6 km (1 mile) below the mean radius of the planet, while the highest peaks, the Maxwell Mountains adjoining Ishtar, rise to over 9.5 km (6 miles).

Clearly the features found on Venus had to be named, and it was laid down that all the chosen names must be female. Thus we have Nightingale, Earhart, Helen, Guinevere and others which are equally familiar. Others were less well-known, such as Quetzalpetlatl, Al-Taymuriyya, Erxleben and Xiao Hong. (In case you are wondering, Quetzalpetlatl is an Aztec fertility goddess, Al-Taymuriyya is an Egyptian authoress, Erxleben is a German scholar and Xiao Hong is a Chinese novelist.) The only masculine name honours the great Scottish mathematician James Clerk Maxwell. This particular name had already been given before the official decree was ratified!

Why is Venus so unlike the Earth? The answer can only lie in its lesser distance from the Sun. It seems that in the early days of the Solar System the Sun was less luminous than it is now, in which case Venus and the Earth may have started to evolve along the same lines, but when the Sun became more powerful the whole situation changed. Earth, at 150 million km (93 million miles), was just out of harm's way, but Venus, at only 108 million km (67 million miles), was not: the oceans boiled away, the carbonates were driven out of the rocks, and in a relatively short time on the cosmical scale Venus was transformed from a potentially life-bearing world into the inferno of today. Any life which had managed to gain a foothold there was ruthlessly destroyed. It is sobering to reflect that if the Earth had been 32 million km (20 million miles) closer to the Sun, the same thing would have happened here.

A new probe, Venus Express, is now orbiting the planet and sending back excellent data. Other unmanned probes will follow, and even land there, but certainly there will be no manned expeditions. It has been suggested that we might eventually be able to 'seed' the atmosphere, breaking up the carbon dioxide and sulphuric acid and releasing free oxygen, but this is quite beyond our powers at the moment. In fact, the more we learn about Venus the less inviting it becomes, and it is rather a relief to turn our attention to the red planet Mars.

Mars is less unlike the Earth than any other planet (see Figure 9.3 and Plate 4). It is much smaller and less massive, and its low escape velocity (just over 5 km/3 miles per second) means that it has lost most of any atmosphere it may once have had; what remains is mainly carbon dioxide, but there is not enough of it to produce a Venus-like greenhouse effect, and Mars is a chilly place. At noon on the equator, in summer, a thermometer would rise well above freezing point, but the nights are bitterly cold – far colder than a polar night on Earth – because the tenuous atmosphere is very poor at retaining heat. The surface pressure is below 10 millibars everywhere, which again corresponds to what we usually call a fairly good laboratory vacuum. Certainly there is no chance that future colonists will be able to walk about in the open with only the protection of a warm suit and an oxygen mask.

The axial rotation period is about half an hour longer than ours, so that the 'year' contains 668 Martian days or 'sols'. The tilt of the axis is almost exactly the same as that of the Earth, so that the seasons are of the same general type. There is, however, one important difference. Mars has an orbit which is appreciably eccentric, and the distance from the Sun ranges between over 240 million km (150 million miles) and less than 209 million km (130 million miles). As with Earth, southern summer occurs at perihelion, so that the southern summers are shorter and hotter than those of the north, while the winters are longer and colder.

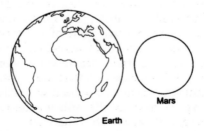

Figure 9.3 Mars and Earth compared

The effect is very marked because there is no ocean to stabilize the temperature – and indeed there can be no liquid water on Mars, because the atmospheric pressure is too low; water would promptly evaporate (except just possibly at the very bottom of the deepest basins).

Telescopes show a red surface, with dark patches and white polar caps. The caps are composed of water ice, plus some carbon dioxide ice, and they are thick – there is no shortage of H_2O on Mars! Predictably, the caps wax and wane with the Martian seasons; at their greatest extent they are very large and prominent, while at their minimum they are very small. The southern cap shows the greater variations in size because of the more extreme climate in the southern hemisphere.

The dark areas were once thought to be seas, but it soon became clear that this could not be so. Next, it was assumed that the dark regions were old sea-beds filled with primitive vegetation. Certainly they are permanent, and the most conspicuous of them, the V-shaped marking now known as the Syrtis Major, was recorded by Christiaan Huygens as long ago as 1659. Maps were drawn up during the nineteenth century, and various systems of nomenclature were introduced; at first the names were given, lunar-style, in honour of famous scientists, but in 1877 Schiaparelli drew a new map and proposed a new system, so that the V-shaped feature seen by Huygens and later christened the Kaiser Sea was renamed Syrtis Major. A small telescope will show it when Mars is well placed – that is to say, a month or two to either side of opposition. At times there are dust-storms which cover the whole planet and hide the surface features, but for most of the time the atmosphere is transparent, though it does contain clouds made up of high-altitude ice crystals.

It was Schiaparelli who made the first detailed drawings of the linear features which he called 'canali', or channels. Inevitably this was translated into English as 'canals', and led to the suggestion that we might be seeing artificial waterways, built by the local inhabitants to form a planet-wide irrigation system. Percival Lowell, founder of the great observatory at Flagstaff in Arizona, was firm in his conviction that Mars was the home of a brilliant civilization. Alas, we now know that the canals do not exist; they were tricks of the eye, and have no basis in reality. The only planet which has canals on its surface is our own Earth. On the other hand it is quite permissible to describe the ochre areas on Mars as deserts; this is precisely what they are, even though they are not 'sandy', and owe their colour to reddish minerals. Mars is, in fact, a rusty kind of place.

Before the Space Age it was still thought that Mars might be reasonably friendly, with a nitrogen atmosphere thick enough to be useful. It was only with the flight of Mariner 4, in 1965, that our ideas changed. The dark areas turned out to be nothing more than albedo features, where the reddish material had been blown away by winds in the thin atmosphere to reveal the darker surface beneath – and instead of being no more than gently undulating, the surface was seen to be heavily cratered, so that Mars was obviously much more 'lunar' than terrestrial. Later space-craft such as Mariner 9, which entered a closed path around Mars in 1971, showed that there are mountains, canyons, valleys, ridges and also massive volcanoes, one of which, Olympus Mons, towers to three times the height of our Everest, and is crowned by an 80-km (50-mile) caldera. It had been shown on earlier maps, and known as Nix Olympica, the 'Olympic Snow', but nobody had any idea of its true nature.

The volcanoes are of special interest. It is generally believed that they are extinct, but we cannot be quite sure, and there is no doubt that there was violent activity in the past. The volcanoes are found in various parts of Mars, but most notably in the area of the so-called Tharsis Bulge, where there is a whole line of giant volcanoes besides which our Hawaiian peaks would seem very puny. There are also some deep basins, one of which, Hellas, measures 2204 by 1802 km (1370 by 1120 miles). It was once believed to be a snow-covered plateau, because it can sometimes appear so bright that it can easily be mistaken for an extra polar cap.

The Mariner pictures also revealed huge canyon systems, dwarfing our much-vaunted Grand Canyon of Colorado, and there were also features which looked so much like dried-up riverbeds that there seems to be no other explanation. (I remember a comment from a celebrated American astronomer: 'If it looks like a duck, waddles like a duck and quacks like a duck, then maybe it *is* a duck.') In this case there must have been running water on Mars in the distant past, which means that the atmosphere must then have been much denser than it is today.

Yet how old are the riverbeds? It is usually assumed that they are very ancient, but there seems to be a surprising lack of erosion; we know that there is a great deal of dust in the Martian atmosphere, and dust is highly abrasive. There are two alternative explanations, though neither is officially favoured.

At present the tilt of the Martian axis of rotation is almost the same as ours (24 degrees to the perpendicular), but it changes in a cycle of about 100 000 years, and at times the inclination may be as much as 35 degrees or as little as 14 degrees. Suppose that at one stage in the precessional cycle the south pole is turned sunward, and that the ices there evaporate, temporarily thickening the atmosphere and making rainfall possible? Again, can it be that Mars goes through periods of intense vulcanism, so that gases and vapours are sent out in sufficient quantity to increase the atmospheric density? In any case, the dry riverbeds are very much in evidence.

If conditions on Mars were once much more pleasant than they are today, it is possible that life might have started to evolve there, dying out when the climate became unsuitable. If so we might expect to find fossils, albeit of a primitive kind. And the question of life was very much in astronomers' minds when, in 1975, the two Viking probes were launched. Their aim was to make controlled landings and then search for any trace of Martian life, past or present. Each vehicle was made up of two parts: an orbiter, and a lander. The two components crossed space together, and when in Martian orbit they were separated, so that the orbiter could continue circling the planet and also act as a relay for the lander. The two Vikings were perfect twins, and in the event both worked faultlessly.

It was an ambitious project. If a lander came down on a rock, for example, it would be unable to communicate, but luckily this did not happen. Of course the landing sites had been surveyed as carefully as possible; before touch-down, the orbiter of Viking 1 sent back a picture of a rock with light and shade effects which gave an uncanny resemblance to a human face. (It is only too easy to tell the reactions of the various brands of eccentrics!)

The Vikings came down gently, No. 1 in the 'golden plain' of Chryse and No. 2 in the more northerly plain inappropriately named Utopia. Material was scooped up from the surface, drawn into the space-craft and analysed, after which the results were transmitted back to Earth. The results showed that there is something decidedly curious about Martian chemistry, but no definite signs of organic activity were found. The next successful lander, Pathfinder, reached Mars in July 1997 (there had been several failures during the interim, and even now the Russians have had no real success with any of their Mars probes, despite their brilliant achievements with Venus). This time there was no attempt to make a 'soft' landing. Pathfinder was encased in airbags and on impact it literally bounced in the manner of a beach-ball. It came to rest in an upright position and subsequently released a tiny 'rover', Sojourner, which crawled around the surface, guided by its controllers many millions of kilometres away. The chosen sites was Ares Vallis, which had been carved by a violent flood in the remote past. Sojourner itself was only just over 0.6 m (2 ft) long, with six wheels. It was amazingly effective and carried out analyses of the rocks, most of which proved to be volcanic. There was also evidence of layering or bedding, suggesting a sedimentary origin.

Mars Global Surveyor entered Mars' orbit in September 1997, and began a long programme of mapping and analysis. One particularly interesting feature was a small crater inside a larger formation, Newton, where there were numerous gullies running from the top of the crater wall; it looked as though they had been cut by running water, and it is possible that liquid burst out from below, eroded the gullies and then froze and evaporated. If so, then there might be water not so far below the Martian surface, and the existence of lowly life-forms cannot be ruled out.

Little can be said about another American probe, Mars Climate Orbiter, which was lost in 1999 because the two teams involved in the approach manoeuvre were working in different units of measurement – one was working in Imperial units and the other in metric, a slight mistake which cost us $125 million – and Mars Polar Lander simply 'went silent' for reasons unknown. Another failure was Britain's Beagle 2, which was designed to make a deliberate search for life. Up to the very last moment everything seemed to be going well, and presumably the landing was made on schedule – but whether Beagle came down intact we do not know because nothing was ever heard from it after arrival.

Yet there were plenty of successes in the early years of the new century: the orbiter Mars Odyssey (2001) worked well, and Mars Express (June 2003) and Mars Reconnaissance Orbiter (April 2005) were still transmitting in 2008.

However, all these pale in comparison with NASA's two rovers, Spirit and Opportunity, which have exceeded all the hopes of their makers. Each landed safely by the use of both airbags and parachutes; both have moved around, sending back images plus a vast amount of general information; and both have remained active long after the end of their scheduled lifetimes of a mere 90 days. Both rovers were launched in June 2003. Spirit came down first, on 4 January 2004, in a crater named Gusev, selected because it is believed to be an ancient lake which has long since dried up. Spirit has confirmed this; it has explored the crater floor, and climbed some modest hills, taking soil analyses all the time. It has had its problems; at one stage its upper parts were covered with 'dust' which threatened to stop its transmissions, but an obliging gust of wind blew the dust away. Opportunity landed on 25 January 2004 on the opposite side of Mars; the area was that of the Meridiani Terra. Opportunity actually touched down inside a small crater which has been named Eagle. During its trek across the landscape it has sent back detailed images, and made regular analyses of surface material. It has even risked making its way on to the floor of a much larger crater, Victoria. As I write these words (January 2008), the saga of the rovers is by no means over; they were not themselves designed to search for life, but they have certainly shown the way, and other probes will follow regularly.

Though Mars is closer than any other planet apart from Venus, it is not too easy to 'observe' in detail, because a small telescope will not show much apart from the polar caps and the main dark areas. However, owners of more powerful instruments can do useful work in monitoring the changes in the polar caps and, above all, tracking clouds and dust-storms.

Mars has two satellites, both discovered by Asaph Hall in 1877; they were named Phobos and Deimos (Fear and Terror) after two attendants of the mythological war-god. Both are tiny, and irregular in shape; Phobos has a longest diameter of less than 32 km (20 miles), Deimos less than 16 km (10 miles). Viking and Mariner pictures show that both are darkish and cratered, and indeed the largest crater on Phobos (named Stickney; this was the maiden name of Hall's wife) has one-quarter the diameter of the satellite itself, so that if it had been formed by a meteorite impact Phobos would have been in grave danger of being shattered into fragments.

Phobos and Deimos are very close to Mars. Indeed, Phobos moves at only about 5790 km (3600 miles) above the planet's surface, and it has an orbital period of 7 hours 39 minutes, so that to a Martian observer it would rise in the west, cross the sky at a breakneck speed and set in the east only 4$\frac{1}{2}$ hours later. It would be of little use as a source of illumination at night, and it would frequently be eclipsed by the shadow of Mars; it would transit the Sun over 1000 times per year, taking 19 seconds to cross the disk. Calculations indicate that Phobos is in an unstable orbit. It is spiralling very slowly downward, and may well crash on to the planet in 40 million years' time.

Deimos is further out, and would remain above the Martian horizon for two and a half sols consecutively, but its phases would be none too easy to see with the naked eye, and the total light sent to the planet would be little greater than that which Venus sends to us. Unlike Phobos, it is in a stable orbit.

There seems little doubt that both Phobos and Deimos are ex-asteroids. There is a rather amusing tale linked with them. In the famous *Gulliver's Travels*, Dean Swift described how the astronomers of the flying island of Laputa had detected two Martian satellites, one of which had an orbital period shorter than the planet's 'day'; inevitably this was seized upon by later eccentrics to claim that Swift knew all about them – even though at the time there was no telescope even remotely powerful enough to show them. In fact, there is a simple explanation. Venus has no moon, the Earth has one, and Jupiter – much further away from the Sun – has four; so how can Mars possibly manage with less than two?

Plans are already being worked for sending the first manned expedition to Mars, and it has been suggested that the date could be as early as 2030. This may be optimistic, but at least we may be confident that Mars will be our first planetary target, and it is quite likely that 'the first man on Mars' has already been born. Travel there now seems much less fantastic than reaching the Moon did at the end of the Second World War.

Meanwhile, Mars is there for our inspection. Look at it through a telescope when you have the chance; you will see the white polar ices, the dark areas and the ochre deserts, and you may also see clouds – though, regrettably, you will see no canals.

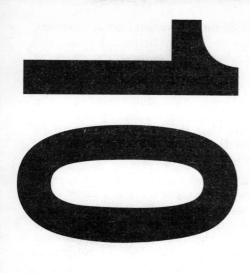

10 the outer planets

In this chapter you will learn:
- about the outer members of the Solar System
- that Jupiter and Saturn are gas-giants
- that Uranus and Neptune may be better called ice-giants.

Far beyond Mars, and well beyond the main belt of asteroids, we come to the four giants: Jupiter, Saturn, Uranus and Neptune. They are often discussed together, and in many respects this is logical enough, but the Jupiter/Saturn pair is very different from the Uranus/Neptune pair, and each planet has its own unique characteristics. Jupiter is the senior member, and it has even been said that the Solar System is made up of the Sun, Jupiter, and assorted débris, but the others are not to be despised; even Uranus, the least massive of the four, could balance over 14 Earths.

Jupiter (see Plate 8), so aptly named after the ruler of Olympus, is a real giant by planetary standards, but it is not a 'failed star', as has been suggested. A normal star shines because of nuclear reactions going on deep inside it, as we have already noted in the case of the Sun. To trigger off the hydrogen-into-helium reaction, the temperature must rise to around 10 000 000° C, but Jupiter, vast though it is, has only a little more than one-thousandth the mass of the Sun, and the core temperature cannot be more than 50 000°, while 30 000° is probably a better estimate. This is far too low for nuclear reactions to begin; to rank as a star, the mass of Jupiter would have to be increased by a factor of ten.

Yet Jupiter does send out more energy than it would if it relied entirely upon what it receives from the Sun. There are two possible explanations for this. The globe could be very slowly shrinking, releasing gravitational energy in the process; the rate of contraction would be far too slow to be measured from Earth. However, it is more likely that Jupiter has simply not had enough time to lose all the heat which it acquired during its formation from the solar nebula, between 4000 and 5000 million years ago. The outer clouds are very cold indeed, at a temperature of around −150° C, so that there is no question of Jupiter sending out enough heat to warm its satellite system.

Look at Jupiter through a telescope – even a small one – and you will see a yellowish, flattened globe crossed by dark streaks which are called cloud belts (see Figure 10.1). There is no mystery about the flattening. Though Jupiter's 'year' is almost 12 times as long as ours, it is a very quick spinner, and a Jovian 'day' is less than ten hours long. As with the Sun, we have differential rotation. The mean equatorial period is 9 hours 50 minutes 30 seconds, while that of the rest of the planet is around five minutes longer, but various discrete features have periods of their own, and drift around in longitude. The equatorial regions bulge out; the diameter as measured through the equator is

Figure 10.1 Jupiter's flattened shape

143 877 km (89 420 miles), while the polar diameter is a mere 133 708 km (83 100 miles). (*En passant*, the difference between the equatorial and polar diameters of the Earth is only 43 km (27 miles), but of course the Earth is a rigid body.)

Obviously we can see only the uppermost clouds of Jupiter, and when we try to investigate the lower layers we have to depend upon theory; it was only in the 1920s that it was finally established that Jupiter is not a midget sun. According to the current models there is a hot core, rich in iron and silicates, surrounded by deep layers of liquid hydrogen (see Figure 10.2). Near the core, the hydrogen is so compressed that it takes on the characteristics of a metal, while higher up we come to normal molecular liquid hydrogen. Above this comes the 'atmosphere', at least 965 km (600 miles) deep, though with a planet such as Jupiter it is not easy to decide just where the 'atmosphere' ends and the main body of the planet begins; the transition must be gradual. It seems that the uppermost clouds are of ammonia ice. Below this comes a layer of ammonium hydrosulphide, a foul-smelling substance made up of ammonia and hydrogen sulphide; deeper still there is a layer of water ice crystals or liquid water droplets, at a pleasantly comfortable temperature. Certainly the main atmosphere consists mainly of hydrogen, which accounts for more than 80 per cent of the total, and there is spectroscopic evidence of hydrogen compounds such as ammonia and methane. Most of the rest of the Jovian atmosphere is composed of helium.

The famous dark belts dominate the scene. Generally there are several belts prominent enough to be seen with a very small telescope; Figure 10.3 shows the accepted nomenclature. The North Equatorial Belt (NEB) is always very conspicuous; the

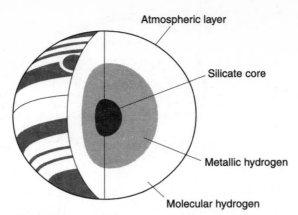

Figure 10.2 Structure of Jupiter

South Equatorial Belt (SEB) is much more variable, and has been known to disappear briefly. The region between these two belts has the quickest rotation, and is termed System I, while the rest of the planet is included in System II. As well as the belts and bright zones there are spots, wisps and festoons, most of which show marked changes even over a few rotations of the planet. The most famous feature is the Great Red Spot (R.S. in Figure 10.3) which was first recorded during the seventeenth century and has been on view for most of the time ever since; at its greatest extent it is 48 270 km long by 9654 km wide (30 000 miles by 6000 miles), so that its surface area is greater than that of the Earth. At times it is strongly red, though on occasion it may look grey. Even when at its most obscure its position can usually be found from the Red Spot Hollow, which affects the boundary of the adjacent belt. The Spot drifts along in longitude, but its latitude does not vary much.

Before the space missions nobody knew quite what to make of the Great Red Spot. The idea of a red-hot volcano poking out above the cloud-deck was abandoned long ago, but it was often believed that the Spot was a solid or semi-solid body floating in the atmosphere, vanishing for a while when it sank and was covered up. Thanks to the probes, we now know the Spot to be a whirling storm – a phenomenon of Jovian 'weather', if you like – spinning around anti-clockwise in a period of 11 days, and having a profound effect upon the whole of that part of Jupiter. The colour is still something of a mystery, but phosphorus may be responsible, and certainly there is a rich

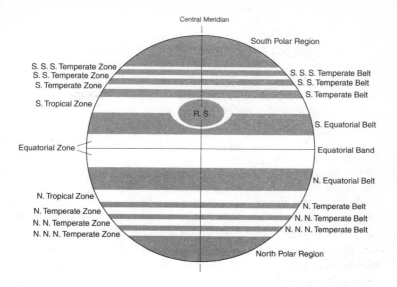

Central Meridian

South Polar Region

S. S. S. Temperate Zone
S. S. Temperate Zone
S. Temperate Zone

S. S. S. Temperate Belt
S. S. Temperate Belt
S. Temperate Belt

S. Tropical Zone

R. S.

S. Equatorial Belt

Equatorial Zone

Equatorial Band

N. Equatorial Belt

N. Tropical Zone

N. Temperate Zone
N. N. Temperate Zone
N. N. N. Temperate Zone

N. Temperate Belt
N. N. Temperate Belt
N. N. N. Temperate Belt

North Polar Region

Figure 10.3 Jovian nomenclature

Jovian chemistry which gives rise to vivid colours not found on any of the other giant planets.

However, the Red Spot is no longer unique. In March 2000 two white ovals merged to form a larger oval. In February 2006 this new oval became red; it too lies in the South Temperate Belt, following the Great Red Spot. It is about half the size of the Red Spot, and must be of the same nature. How long it will persist remains to be seen.

There was a dramatic event in July 1994, when a comet – Shoemaker–Levy 9 – crashed on to Jupiter. The comet is so named because it was the ninth discovery made jointly by two professional astronomers (Eugene and Carolyn Shoemaker) and one amateur (David Levy). Its nucleus had already been disrupted during an earlier encounter with the planet, and had been torn into a chain of over 20 fragments, which impacted in turn between 16 and 22 July 1994. Predictably, the astrologers and the end-of-the-worlders had been in full cry, forecasting doom and destruction; astronomers were doubtful whether anything much would be seen. In fact, the results of the impacts were very obvious, and for a while Jupiter assumed a decidedly unfamiliar appearance, with black patches in the far southern hemisphere showing where the fragments had hit. The storms generated were in many cases much larger than the Earth. The full results took years to analyse in full, and gave us new information about the composition of the Jovian clouds. No doubt such impacts have occurred many times in the past, but this is the first time that a cometary suicide has been 'caught in the act' with respect to Jupiter, and it will be strange if we see anything comparable in the near future.

During the 1950s it was found that Jupiter is a source of radio waves, and it was tacitly assumed that there must be a strong magnetic field, but reliable information had to await the missions of the Pioneers and Voyagers. The Pioneers – Nos 10 and 11 in the series – were launched in 1972 and 1973 respectively, and each passed by Jupiter after journeys lasting for the best part of two years, though they were very successful, they were outshone by the Voyagers, which followed a few years later.

Voyager 2 was launched on 20 August 1977, and by-passed Jupiter on 8 July 1979 at a range of 707 960 km (440 000 miles). Voyager 1 began its journey slightly later, on 5 September 1977, but travelled in a more economical path, and made its pass of Jupiter on 5 March 1979 at a minimum distance of only 349 150 km (217 000 miles). This was somewhat risky, because the earlier Pioneers had confirmed that Jupiter is associated with strong zones of radiation – and it is quite definite that any astronaut foolish enough to venture within 160 900 km (100 000 miles) of those innocent-looking clouds would die quickly and unpleasantly from radiation poisoning. Both Voyagers were programmed to pass as quickly as possible over the equatorial regions of the globe, where the danger is at its worst.

Detailed analyses were made of the clouds, and the favoured model of the interior was confirmed as far as possible. It became patently clear that Jupiter is a world in constant turmoil; storms rage unceasingly, with lightning and, no doubt, thunder. Aurorae were recorded on the night side. Moreover, the Voyagers show that Jupiter has rings, though they are dark and obscure and not in the least like the glorious ring-system of Saturn.

The next impact on Jupiter was man-made. The Galileo probe was launched on 18 October 1989 and, after a somewhat circuitous journey, reached Jupiter on 7 December 1995. It was made up of two parts: an orbiter, and an entry probe, which were separated well before arrival. The entry probe plunged into the Jovian clouds at over 170 554 kmh (106 000 mph), and sent back signals for 57 minutes before being destroyed. There were several surprises – mainly the fact that signs of the expected watery layer did not show up. Meanwhile, the orbiting section of Galileo entered a closed path around the planet, and began sending back information not only about Jupiter itself, but about its satellites – which have proved to be even more interesting than had been expected during the pre-Space Age.

Galileo Galilei, with his primitive telescope, observed the four main satellites in 1610, which is why they are always known collectively as the Galileans, though in fact Galileo was not the first to see them telescopically; priority seems to go to his contemporary Simon Marius, and it was Marius who named them Io, Europa, Ganymede and Callisto (which may be why these names were not generally used in pre-Space Age days). Any telescope will show them, and so will good binoculars, while really keen-sighted people can see Ganymede at least with the naked eye; it was recorded by a Chinese star-gazer, Gan De, as long ago as 364 BC.

All the satellites are planetary in size. Ganymede and Callisto are much larger than our Moon, and Ganymede is actually larger than Mercury, though it is less massive and, like the other Galileans, has no appreciable atmosphere, though there is an excessively tenuous 'exosphere' consisting of oxygen, and there is a weak but detectable magnetic field.

The Galileans make up a fascinating family. The images sent back by the space probes show that Io is wildly volcanic, with sulphur and sulphur dioxide volcanoes erupting all the time; it has been said that Io looks rather like an Italian pizza. There are extensive lava-flows and the hot volcanic vents with their calderas filled with black lava erupt at temperatures of hundreds of degrees; the most active volcano, named Loki, produces lava at well over 1000° C, even though the general temperature of the surface is a chilling –140° C. Vast amounts of volcanic material are sent out, the ejected material is spread all around Io's orbit, and the satellite is connected to Jupiter by a powerful electric current. This explains why Io's position in orbit has a marked effect upon the radio emissions from Jupiter. The sulphur dioxide 'atmosphere' is a thousand million times less dense than our own air and, since Io moves well within Jupiter's radiation zones, it may lay claim to being the most lethal world in the Solar System. It is always changing, and the outbreaks there can now be monitored with the Hubble Space Telescope.

Europa is as different from Io as it could possibly be. It has a smooth surface and there is no doubt that the surface material is water ice; there are few impact craters and little vertical relief, but there are criss-crossing low ridges and shallow grooves, making Europa a mapmaker's nightmare. Unexpectedly, Europa is believed to have an ocean of salty water below the ice-crust, though the chances of life there are remote. It may seem strange that Europa is inert while Io is so active; the official theory is that Io's interior is 'flexed' by the changing gravitational pulls of Jupiter and the other Galileans, but Europa is much further away from Jupiter, so that the flexing effects are less.

Ganymede is much less dense than Io or Europa; the surface shows bright and dark regions, with many craters and grooves. Callisto is heavily cratered and there is a distinct chance that it, like Europa, may have an ocean below its icy crust. There are several huge basins, one of which, Valhalla, has a longest diameter of over 1600 km (1000 miles).

Four small satellites – Metis, Adrastea, Amalthea and Thebe – move around Jupiter closer-in than the Galileans; Amalthea, discovered by E. E. Barnard in 1892, is the largest of them, with a diameter of 262 km (163 miles). Beyond the orbit of Callisto come other satellites, of which the largest, Himalia, is 251 km (156 miles) across. Some of these dwarfs move in retrograde orbits and all are certainly captured asteroids rather than bona-fide satellites.

Jupiter is a rewarding object for the user of a small or moderate-sized telescope. The quick spin means that the spots and other markings are carried relatively rapidly across the disk; the shifts are evident even after a few minutes. Timing the moments when the different features reach the central meridian means that the longitudes can be worked out, and rotation periods derived; the central meridian is easy to locate, because of the flattening of the globe, and estimates can be made to an accuracy of less than a minute. Until fairly recent times it was amateur work which gave us our best data about the rotation periods of the Jovian features, and work of this kind is still very useful.

Next in order of distance from the Sun comes Saturn (see Plate 9), arguably the most beautiful object in the entire sky. To the naked eye it looks ordinary enough; it takes the guise of a bright, rather yellowish star, whose 'dull' glare led the ancients to name it after the God of Time (also Jupiter's father, and the previous ruler of Olympus). It takes 29½ years to complete one orbit, and is therefore a slow mover in the sky; once identified, it is easy to find again.

In size and mass, Saturn is second only to Jupiter. Though it is of the same basic nature, there are some important differences, and in particular Saturn's overall density is lower; it is actually less than that of water, so that although the mass is 95 times greater than that of the Earth the surface gravity is less than 1.2 times greater. The surface is gaseous, and hydrogen is (predictably) the main constituent, with most of the rest of the atmosphere being made up of helium. Below the clouds come deep layers of liquid hydrogen, and finally a silicate core not a great deal larger than the Earth. The core temperature may be as high as 15 000° C, but the outer clouds are bitterly cold, at around –180° C. Belts and bright zones are visible with a small telescope, but are less pronounced than those of Jupiter, and there is nothing to match Jupiter's Great Red Spot.

There is a magnetic field, much stronger than that of the Earth though not nearly so powerful as that of Jupiter; and Saturn is a quick spinner, which means that its yellowish globe is very obviously flattened. The axial inclination is almost 27 degrees to the perpendicular, much greater than with Jupiter, and the magnetic and rotational axes are almost coincident.

Occasional outbreaks are seen in the form of white spots. There have been two of special note during the present century, one in 1933 (discovered by a skilful amateur astronomer named W. T. Hay), and the other in 1990. Both of these became very conspicuous, and lasted for some time. Undoubtedly they were due to material gushing up from below cloud-level.

Four space-craft have so far encountered Saturn. The first of these was Pioneer 11, which had already passed by Jupiter (in 1974) and was sent on to Saturn more or less as an afterthought. It sent back good pictures, but better results have come from the Voyagers, both of which were amazingly successful. Voyager 1 approached Saturn to within a distance of 125 502 km (78 000 miles) on 12 November 1980, and Voyager 2 passed by at a mere 101 370 km (63 000 miles) on 25 August 1981. At least the radiation danger was known to be much less than with Jupiter, because Saturn's zones are much weaker.

Excellent views of the globe were obtained, but of course the main glory of Saturn lies with its rings, which are made up of vast numbers of particles of water ice, all spinning around the planet in the manner of moonlets; no solid or liquid ring could possibly exist, since it would promptly be broken up by Saturn's powerful pull (see Figure 10.4). There are two main rings, A and B, separated by a gap known as Cassini's Division in honour of the

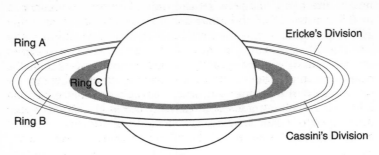

Figure 10.4 Saturn's ring system

Italian astronomer who discovered it, from the Paris Observatory, in 1675. (Giovanni Cassini had his problems at Paris. The French King insisted that the new Observatory should be architecturally attractive rather than functional, so that the sky was largely masked by inconvenient turrets and other structures. Eventually, in desperation, Cassini took his telescopes out into the Observatory grounds, and used them from there!) The outer ring, A, shows a narrower division which was first definitely recorded by J. F. Encke in 1837. Between the main rings and the planet there is a much dimmer, semi-transparent ring, known officially as Ring C but more commonly as the Crêpe or Dusky Ring.

Naturally, the rings are circular (or virtually so), but from Earth they appear elliptical, owing to the angle at which we see them. When best displayed, as in the early 2000s, they are magnificent even in a small telescope. The Cassini Division is easy to see, and a 15-cm (6-inch) telescope will bring out the Encke Division as well as the Crêpe Ring. However, when the system is edgewise-on, as in 1995, the rings appear only as a thin line of light, and when the Earth and the Sun are passing through the main plane only giant telescopes will show the rings at all. The system measures 271 920 km (169 000 miles) from end to end, but the thickness cannot be more than 0.8 km (½ mile).

Voyager results showed that the ring system is much more complex than had been expected. There are in fact many hundreds of minor ringlets and narrow divisions, and there are ringlets even in the Cassini and Encke Divisions; the Encke Division even contains a tiny satellite, now named Pan. The brightest ring, B, shows strange dark radial spokes, presumably elevated away from the main plane by magnetic forces. The Voyagers also detected several faint new rings outside the main system, all of which are virtually impossible to see with Earth-based telescopes. The next Saturn probe, Cassini, was launched on 15 October 1997 and it has been orbiting Saturn since 1 July 2004. (En route it flew past Jupiter – 26 December 2000 – and sent back excellent images and data.) Cassini carried instruments which have greatly extended our knowledge of the rings, and, more spectacularly, it carried a lander – Huygens – which came down gently on Titan, Saturn's major satellite, on 25 December 2004. This was arguably the most difficult space operation to date. Cassini has since been 'touring' around Saturn, and has surveyed all the main satellites at close range.

Saturn has a wealth of satellites but only one – Titan – is large; seven are of what may be termed medium size, while the rest are true midgets. Titan is unique among planetary satellites in having a dense atmosphere, which Voyager 1 found to be made up chiefly of nitrogen, together with a good deal of methane. Voyager could tell us no more, but Cassini and Huygens told us at last what Titan is really like.

Huygens came down upon a spongy solid surface; there was not much surface relief, but there were channels clearly cut by running liquid – not water, of course, but liquid methane. There must be frequent methane rain, and there may be a steady methane drizzle. Later images from the orbiting Cassini showed what are certainly large lakes filled with hydrocarbons. The Lake District of Titan may not be a cheerful place, but future explorers will find it fascinating. Life seems to be unlikely under these intensely cold, hostile conditions.

All the medium-sized satellites proved to be surprising in their different ways. Mimas has one huge crater (Herschel) which has a diameter one-third that of the satellite itself; Tethys and Dione may have oceans below their icy crusts; Rhea has a battered surface which looks very ancient. Hyperion is shaped rather like a hamburger, with a longest diameter of 377 km (234 miles); its orbital period is over 21 days, but its rotational period is 'chaotic', and is at present about 13 days.

Enceladus is remarkable inasmuch as it is active. Fountains of icy particles are sent out from its polar vents, and underground water is a virtual certainty; this was the last thing that we expected with a world as small as Enceladus. The other real oddity is Iapetus, almost 1448 km (900 miles) across, which is much brighter when west of Saturn than when it is to the east. This is because the rotational period is the same as the orbital period (79 days) and the two hemispheres are of unequal reflecting power; one is bright and icy, the other is as dark as a blackboard. During western elongations it is always the brighter hemisphere which faces us, and Iapetus can then be seen with a small telescope, when to the east it is much more elusive. Because the mean density of the globe is low, we may be sure that a good part of it is composed of ice, and that the dark areas are merely stained in some ways. Cassini has shown that both the bright and the dark regions are cratered – and that a towering mountain ridge, over 8 km (5 miles) above the general surface level, runs along the equator, which does not mark the line of demarcation between the bright and dark areas. Frankly, Iapetus is an enigma. The much smaller Phoebe, 13 million km (8 million miles) from Saturn, is (predictably) cratered; it has retrograde motion and, like other members of the swarm of outer attendants, is certainly a captured body rather than a bona-fide satellite.

Of the small inner satellites, Janus and Epimetheus are of interest because they move in almost identical paths, and there seems no doubt that they are parts of a former larger satellite which broke up; each is very irregular in shape, as the Voyager imaged showed. Periodically they approach each other and exchange orbits, rather as though playing a game of cosmic musical chairs.

Saturn is always worth observing, partly for its beauty and partly because there is always the chance of finding a new spectacular white spot. It is also convenient, inasmuch as it will usually stand a high magnification well. A 15-cm (6-inch) telescope will show Rhea, Iapetus (when at its best) and Dione and Tethys as well as Titan, which is of course an easy object and has even been glimpsed with good binoculars.

Beyond Saturn move the two outer giants, discovered in near-modern times: Uranus in 1781, by Sir William Herschel, and Neptune in 1846, by Johann Galle and Heinrich D'Arrest, who were searching in a position given by the French mathematician Urbain Le Verrier. The two are near-twins in size and mass. Both are around 50 000 km (31 000 miles) in diameter; Neptune is very slightly the smaller, but appreciably the denser and more massive. Both seem to be made up largely of 'ices', with hydrogen-rich gaseous atmospheres which form the surfaces we can see.

In some ways the two are genuinely alike, but in other respects they are very different. In particular Neptune has a powerful internal heat source, while Uranus has not; Neptune's rotational axis is inclined at an angle of less than 30 degrees, while with Uranus the tilt is 98 degrees – more than a right angle, so that

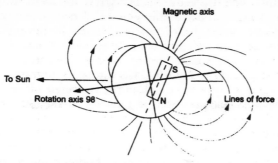

Figure 10.5 Magnetic axis of Uranus

there are times when one or other of the poles is facing the Sun (see Figure 10.5), and the Uranian calendar can only be described as weird. Even the Voyager 2 pictures showed very little detail on Uranus, but Neptune proved to be much more dynamic, with one striking feature, now called the Great Dark Spot, and a beautiful blue surface together with high-altitude clouds made up of methane crystals. Both have obscure ring systems, made up of dark particles and therefore more like the system of Jupiter than that of Saturn, and both are sources of radio emission, though their magnetic axes are well away from the rotational axes and do not even pass through the centres of the globes.

The satellite systems are different, too. Uranus has over 20 attendants, of which only 5 were known before the Voyager 2 mission; even Titania, the largest of them, is less than 1609 km (1000 miles) in diameter. All are icy, and Miranda, the smallest of the 'original five', has an extraordinarily varied surface with plains, craters, icy cliffs and large enclosures or 'coronae' which were nicknamed race-tracks! Neptune has one major satellite, Triton, which was well imaged from Voyager 2, and is unique in having pink nitrogen snow at its pole, with active nitrogen geysers squirting out from below the surface. Its diameter is 2705 km (1681 miles), smaller than the Moon but larger than any of the satellites of Uranus – and, incidentally, larger than Pluto. It has retrograde movement, and seems to be a formerly independent body which was ensnared by Neptune. The other satellites are very small, and only one of them, Nereid, was known before the Voyager 2 pass; it is less than 240 km (150 miles) across, and has a eccentric orbit around Neptune, more like that of a comet than a satellite.

Neptune's discovery by Galle and D'Arrest, in 1846, was the result of mathematical calculations by Le Verrier, who showed that Uranus was being pulled out of position by the pull of a more distant planet, and predicted its position. Decades later Percival Lowell made calculations of the same kind, and in 1930 Clyde Tombaugh, at the Lowell Observatory in Arizona, found what was – naturally – regarded as the expected planet, very close to the predicted position of Lowell's 'Planet X'. It was named Pluto, and it was thought to be comparable in size with the Earth, and to be the only planetary body moving beyond the orbit of Neptune. In fact, Pluto is smaller than the Moon, and merely one member of a whole swarm of bodies moving around the Sun in those remote regions; this swarm makes up the Kuiper Belt, named after the Dutch astronomer Gerard Kuiper, who predicted its existence. Pluto is not even the largest known KBO (Kuiper Belt Object); it is appreciably smaller than Eris, discovered in 2003, whose diameter may be as much as 2574 km (1600 miles). Since all these are now known as Small Solar System Bodies (SSSBs), I discuss them in Chapter 11.

We must be definite here: Pluto is not a true planet; it never was. So is there another large planet, beyond the orbit of Neptune and beyond the Kuiper Belt – in fact, Lowell's Planet X? This is not impossible, but the odds are against it. If it does exist, it will no doubt be found one day. We must wait and see.

11

minor members of the Solar System

In this chapter you will learn:
- about the minor bodies of the Solar System – asteroids, comets, meteors and meteorites.

In the early 1770s a German astronomer named Johann Titius drew attention to a curious relationship linking the distances of the planets from the Sun. It was popularized by Johann Elert Bode, Director of the Königsberg Observatory, and – rather unfairly – it is generally known as Bode's Law.

Take the numbers 0, 3, 6, 12, 24, 48, 96 and 192, each of which, apart from the first two, is double its predecessor. Add 4 to each. Taking the Earth's distance from the Sun as 10 units, the remaining figures give the mean distances of the planets with fair accuracy – out as far as Saturn, which was the most remote planet known at that time. Here is the table (Table 11.1), to which I have added Uranus and Neptune for the sake of completeness:

Table 11.1 Bode's Law

Planet	Distance by Bode's Law	Actual distance
Mercury	4	3.9
Venus	7	7.2
Earth	10	10
Mars	16	15.2
–	28	–
Jupiter	52	52.0
Saturn	100	95.4
Uranus	196	191.8
Neptune	388	300.7

When Uranus was found, in 1781, the Bode distance was in fair accord with the actual distance. Of course the Law fails badly for Neptune. All in all it now seems that the relationship is sheer coincidence, belonging more or less to the 'take-away-the-number-you-first-thought-of' status. But this was not apparent in Bode's day, and the missing planet corresponding to the Bode 28 began to assume real significance. It ought to be there, but clearly it would have to be small, as otherwise it would have been found long ago.

In 1800 a team of astronomers assembled at Johann Schröter's observatory at Lilienthal, near Bremen. They called themselves the Celestial Police, and agreed to undertake a systematic hunt for the expected planet, so that each member of the team would be responsible for a particular part of the ecliptic. Ironically, they were forestalled. Before they had time to put the scheme into full working order, Giuseppe Piazzi at the Palermo Observatory in Sicily chanced upon a slow-moving object which

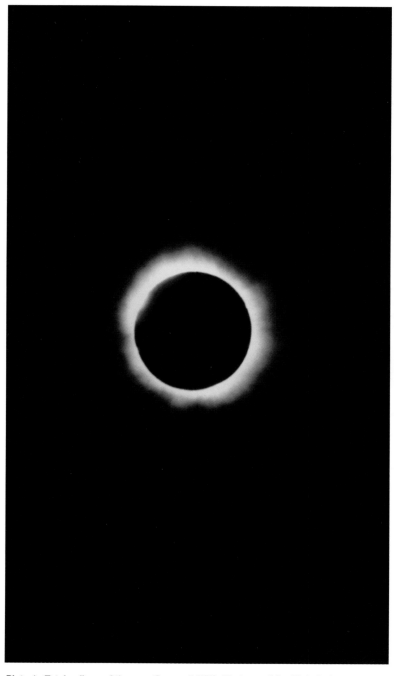

Plate 1   Total eclipse of the sun, Cornwall 1999. Photograph by Chris Doherty.

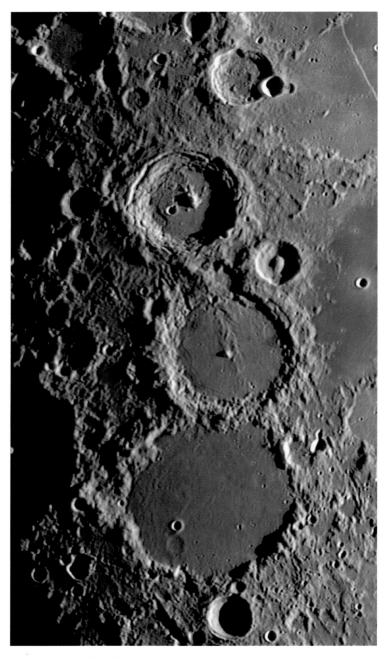

Plate 2 Craters of the Moon. The large crater at the bottom is Ptolemaeus and is 93 miles across. The straight wall is seen to the upper right. Photograph by Bruce Kingsley.

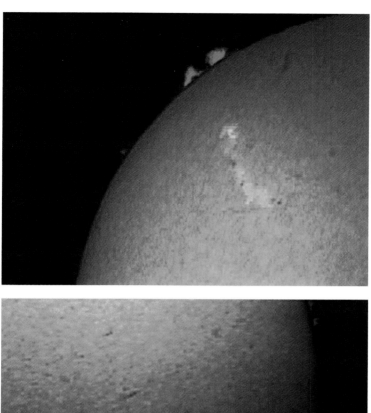

Plate 3 Solar Prominence. Photograph by Jeremy Rundle.

Plate 4   Mars, 2008. The Syrtis Major is central; the north ice cap at the bottom. Photograph by Pete Lawrence.

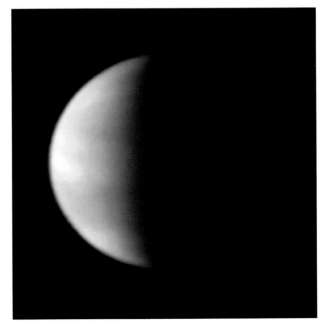

Plate 5   Venus. Photograph by Bruce Kingsley.

Plate 6 The Moon from Apollo 17. The Lunar Rover is shown with Dr Harrison Schmitt setting up the ALSEP (Apollo Lunar Surface Experimental Package).

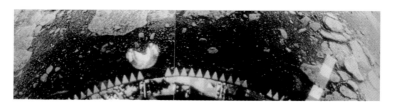

Plate 7 Surface of Venus, from the Russian space-craft Venera 13 (March 1982). Part of the space-craft is shown.

Plate 8   Jupiter, 2008. Photograph by Bruce Kingsley.

Plate 9   Saturn, 2007. Photograph by Pete Lawrence.

Plate 10　The Rosette's Nebula. Photograph by Ian Sharp.

Plate 11 Orion's Sword (M42). Photograph by Ian Sharp.

Plate 12 The Andromeda Galaxy (M31). Photograph by Pete Lawrence.

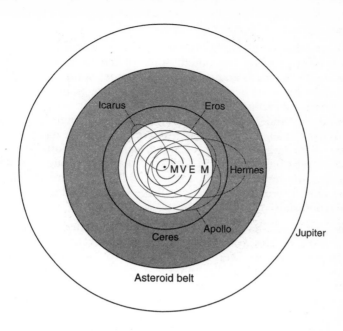

Figure 11.1 The asteroid belt, with some exceptional asteroid orbits

proved to be a small world moving in just the right part of the
Solar System. Its Bode distance was 27.7; Piazzi named it Ceres,
in honour of the patron goddess of Sicily. However, Ceres was
very small, and the 'Police' were not satisfied. Between 1802
and 1808 they discovered three more bodies in the same region:
Pallas, Juno and Vesta. No more seemed to be forthcoming, and
the Police disbanded in 1815 after Schröter's observatory was
sacked by the invading French Army. Then, in 1845, a fifth
'asteroid', Astraea, was found by a German amateur astronomer
named K. L. Hencke; three more followed in 1847, and since
1848 no year has passed without several new asteroid
discoveries. The grand total is now over 100 000, and many
more asteroids have been seen without being followed for long
enough to have their orbits worked out.

Ceres, with a diameter of 940 km (584 miles), is the giant of the
swarm, and of the rest only Vesta and Pallas are more than
480 km (300 miles) across. Only Vesta is ever clearly visible

with the naked eye. Most of the asteroids are tiny, and are not regular in shape. Even Pallas is triaxial, though both Ceres and Vesta appear to be more or less spherical.

All asteroids are assigned numbers and names when their orbits have been properly computed. Some of the names are bizarre; the supply of mythological gods and heroes soon ran out, and we now have such oddities as 518 Halawe (named after an Arabian sweetmeat), 1625 The NORC (named after a computer), and 2309 Mr Spock (honouring a ginger cat which was itself named after the sharp-eared Vulcanian navigator of the fictional starship *Enterprise*). Traditionally, the discoverer of an asteroid is entitled to name it, though the final ratification rests with the International Astronomical Union, and one or two suggested names have been vetoed as being too controversial!

The asteroids have not always been regarded as popular members of the Solar System, since plates taken for quite different reasons were often found to be crawling with unwanted asteroid tracks, and one irritated German went so far as to describe them as 'vermin of the skies'. Yet today they have been rehabilitated, and we regard them as intensely interesting, small though they are. As we have noted, it is now thought that no large planet could form in this region because of the disruptive pull of Jupiter. Moreover, there are well-marked gaps in the main zone where the orbital periods would be exact fractions of that of Jupiter, so that cumulative perturbations keep these gaps swept more or less clean. Asteroids also tend to group in 'families', and collisions must be frequent even today.

Asteroids are of different types; some (such as Ceres) are 'carbonaceous', others (such as Juno) are rich in silicates, while No. 16, Psyche, seems to be almost pure nickel-alloy in composition. Vesta is unusual and is coated with a layer of igneous rock which was originally molten. The first rough map of the surface was obtained in 1995 with the Hubble Space Telescope. It has been suggested that asteroids 1999 (Braille) and 1929 (Kollaa) are broken-off pieces of Vesta, since they seem to have the same exceptional spectra. Ceres is further away, but some details have been seen with the Hubble Telescope, notably what seems to be a large crater or basin which has been appropriately named after Piazzi.

Some asteroids have also been imaged by space-craft. No. 951 Gaspra and 243 Ida were surveyed by the Galileo probe en route for Jupiter and were found to be cratered; 253 Mathilde was encountered by another probe, Shoemaker (named after the famous American planetary geologist Eugene Shoemaker) in 1997. Ida was found to have a tiny satellite, now named Dactyl. Mathilde proved to be as black as charcoal and very irregular; the longest diameter is 66 km (41 miles), but there is one 31-km (19-mile) crater. Fittingly, the Mathilde craters have been named after famous coal-mines. A new probe, Dawn, is now on its way to Vesta and Ceres.

Some asteroids depart from the main swarm and invade the inner part of the Solar System. The first of these to be discovered was 433 Eros, found in 1898, which moves in an orbit crossing that of Mars; at its closest to the Earth it may come within 24 million km (15 million miles) of us. In February 2001 the Shoemaker space-craft made a controlled landing on Eros and discovered craters, boulders and 'ponds', flat surfaces at the bottom of small craters. Eros is shaped rather like a sausage, with a longest diameter of 29 km (18 miles).

There are many other 'near-Earth asteroids' – all of them very small. Several have passed by at distances less than that of the Moon, such as 1994 XM1 (as yet unnamed) which brushed past in December 1994 at less than 112 630 km (70 000 miles). Its diameter was around 9 m (30 ft).

It now seems that these wanderers are much more common than was formerly thought, and several new close-approach asteroids are being discovered yearly. In this case there is always the chance of a collision, and there is some support for the theory that this did happen around 65 million years ago, hurling so much débris into the atmosphere that the Earth's climate was completely changed, with disastrous results for the dinosaurs. Whether this is true or not (and personally I am distinctly dubious), there is no doubt that we are not immune. If we saw such an asteroid on a collision course it is possible that we might be able to divert it by a nuclear missile, but whether we would have sufficient warning is not clear; for example, the 1991 Earth-grazer was discovered only when it had already passed its point of closest approach.

Some asteroids, such as 1566 Icarus and 3200 Phaethon, spend parts of their orbits closer-in than the path of Mercury, so that at perihelion they must be red-hot. Phaethon moves in the same orbit as the Geminid meteor stream, and may well be its 'parent', in which case we have strong confirmation of the link between comets and small asteroids; this is further supported by the strange case of Asteroid 4015, discovered in 1977, which was subsequently found to have been previously seen in 1949 in the guise of a comet, Wilson–Harrington. I will return to this theme later. Meanwhile, note also that there are a few tiny asteroids whose mean distances from the Sun are less than that of the Earth; such is 2100 Ra-Shalom, which ranges between 69 million km and 126 million km (43 million and 78 million miles) from the Sun in a period of 283 days. It is no more than 3.2 km (2 miles) in diameter, and may be considerably less.

There are also asteroids which keep well outside the main swarm. Such are the Trojans, which move in the same orbit as Jupiter, though they keep prudently either 45 degrees ahead of or 45 degrees behind the Giant Planet and are in no danger of being engulfed (though naturally they oscillate for some way to either side of their mean positions in orbit relative to Jupiter). Some of them are over 160 km (100 miles) in diameter, though their remoteness from Earth means that they are very faint.

2060 Chiron, a different kind of body, was discovered in 1977 by Charles Kowal at Palomar. (Do not confuse Chiron with Charon, the satellite of Pluto. It is a pity that the names are so alike, but they are not associated. In mythology Chiron was a wise centaur who taught Jason and the Argonauts, while Charon was the gloomy ferryman who took departed souls across the River Styx on their way to Pluto's kingdom!) By asteroidal standards it is large – over 240 km (150 miles) in diameter, perhaps over 322 km (200 miles) – and it spends most of its time between the orbits of Saturn and Uranus, which at the time of its discovery was the last place where one would have expected to find an asteroid. It reached perihelion in 1996, when its distance from the Sun was 1278 million km (794 million miles); at aphelion it lies at over 2735 million km (1700 million miles). The period is 51 years. It caused a major surprise in 1988, when it was seen to be brightening – not spectacularly, but appreciably. It then developed a sort of 'fuzz', as though an icy layer on the surface starting to evaporate as the temperature rose. This is cometary behaviour, but Chiron is far too big to be a comet, and more likely it is a planetesimal. Calculations show that in 1664 BC it passed within 16 million km (10 million miles) of Saturn; this is not much greater than the distance between Saturn and its outer retrograde satellite Phoebe – and Phoebe and Chiron are of almost the same size.

Other exceptional bodies have since been found. Some have very eccentric orbits which take them from the inner Solar System out into the depths; thus 5365 Damocles has an orbit which crosses those of Mars, Jupiter, Saturn and Uranus, while 5145 Pholus moves from within the orbit of Saturn out to well beyond Neptune in a period of 93 years. It is very red, and its diameter may be from 240 to 322 km (150 to 200 miles).

Let me turn now to the Kuiper Belt. The brightest member of the swarm (though not the largest) is Pluto; like all other KBOs (Kuiper Belt Objects) it has been given an asteroid number, 134340 (Eris is 136199). Pluto has an orbital period of 248 years, and at perihelion it comes within the orbit of Neptune, as it did in 1979 and 1999, but there is no danger of collision because Pluto's orbit is inclined to be ecliptic by as much as 17 degrees. It has one comparatively large satellite, Charon, whose diameter (1190 km/740 miles) is about half that of Pluto itself. Charon's orbital period (6 days/9 hours) is the same as Pluto's axial rotation period, so that the two are 'locked', and an observer on Pluto would see Charon motionless in the sky. There are two much smaller satellites, Nix and Hydra. Pluto is of the fourteenth magnitude, and by no means difficult to see, but detail on its disk is beyond the range of any but the most powerful telescopes, and even they can do no more than make out a few bright and darker areas. Pluto has an extensive though immensely rarefied atmosphere, but as the planet moves out to its next aphelion, in 2113, the temperatue will become so low that the atmosphere may freeze out on to the surface. A NASA space-craft, New Horizons, is now on its way to Pluto, and should make its fly-by in 2015, while the atmosphere is still gaseous.

We know even less about the other KBOs. Eris seems to be 27 per cent more massive than Pluto, but is further away (over 9650 million km/6000 million miles; period 557 years) and has one known satellite, Dysnomia. Another notable KBO is 5000 Quaoar which, like Charon, has shown signs of comet-like activity, even though its diameter, over 1126 km (700 miles), is much too great for it to be classed as any kind of comet.

There are also the Scattered Disk objects, which range out to great distances. 90377 Sedna, discovered in 2003, has a period of 12 059 years, and at aphelion moves out to a distance of about 144 810 million km (90 000 million miles). It is no more than 1930 km (1200 miles) across, so it will be too faint for us to see at all. We were fortunate in catching it near perihelion. No doubt many other Sedna-type objects exist.

Turning now to comets, which present problems of their own (I once described a comet as being 'the nearest approach to nothing that can still be anything'.) They are often called 'dirty snowballs' and this is by no means a bad description. A brilliant comet, with a long tail stretching across the sky, may appear imposing by any standards, but is not nearly so important as it may seem. The only substantial part of a comet is its nucleus, which is made up of a mixture of ice and 'rubble', and is never more than a few kilometres across. The average comet travels in an elliptical orbit, and when far away from the Sun is simply a lump of inert material, but when it draws inward, and the temperature rises, the ices in the nucleus start to evaporate, so that the comet develops a head or coma and perhaps tails. When the comet has passed perihelion, and has drawn back into the cold regions far from the Sun, the head and tails disappear, and the comet reverts to its dormant state.

Tails are of two types: gas (ion) and dust. Both are produced by the effects of the Sun. Light-pressure drives tenuous material outward to produce a straight gas-tail; the solar wind particles also drive material outward to make a curved dust-tail. Both types always point more or less away from the Sun, so that when a comet is moving outward after passing through perihelion it travels tail-first (see Figure 11.2). Of course, many small comets never produce tails of either kind, and some of them look like nothing more than slightly 'fuzzy' stars.

It has always been thought that comets come from the so-called Oort Cloud, a collection of icy objects moving around the Sun at a distance of at least a light-year. (The name honours the Dutch astronomer J. H. Oort, who became the world's leading authority on cometary phenomena.) It was assumed that if a comet was perturbed for any reason, perhaps by a passing star or even by a remote solar planet, it would start to fall inward towards the Sun, and after a journey lasting for many thousands of years would come within our range. One of several things might then happen to it. It might simply swing past the Sun and return to the Oort Cloud, not to return for a very long time. It might encounter a planet (usually Jupiter) and be put into a parabolic orbit, so that it would never return at all; this was the fate of a fairly bright comet (Arend–Roland) seen in the spring of 1957. It might be put into a short-period path with a period of only a few years or a few decades – or it might even fall into the Sun or hit a planet; remember that in July 1994 we witnessed the destruction of a comet, Shoemaker–Levy 9, which committed hara-kiri by impacting Jupiter. However, even though brilliant comets come from the Oort Cloud, the short-

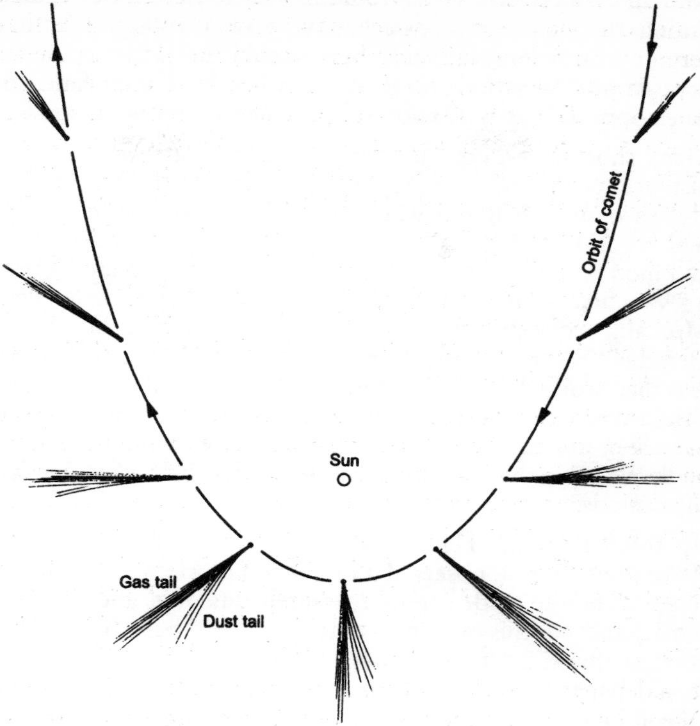

Figure 11.2 Behaviour of a comet's tail

period comets seem to swing in from the Kuiper Belt. Certainly comets are very primitive bodies.

Comets are named after their discoverer or discoverers or, occasionally, after the mathematician who first computed the orbit. No year now passes without its quota of new discoveries – some comets seen for the first time, and others which mark the returns of known periodical comets.

Each time a comet passes through perihelion it loses some of its mass, since material is evaporated to produce the head and tail; this means that by cosmical standards a comet's lifetime is short, and after only a few tens of thousands of years it loses all its volatiles. As we have noted, Comet Wilson–Harrington of 1949 is now purely asteroidal in appearance and has even been given an asteroid number. Short-period comets have to a great extent wasted away, and few of them become bright enough to be seen with the naked eye; several comets which were seen regularly during the nineteenth century have now vanished, and have certainly disintegrated. Such, for example, were the comets of Brorsen and Westphal, while in 1926 Ensor's Comet faded out when approaching perihelion and was never seen again.

Many short-period comets have their aphelia at about the same distance as the orbit of Jupiter. They are old friends, and we always know when and where to expect them; Encke's Comet, with a period of 3.3 years, has now been seen at over 50 returns, and modern telescopes can follow it all the way around its orbit. It was first seen in 1786 by the French astronomer Pierre Méchain; recovered in 1795 by Caroline Herschel, Sir William Herschel's sister; seen again in 1805 and in 1818, and the orbit was then worked out by J. F. Encke, who predicted that the comet would come back in 1822. It duly did so, and since then it has been seen at every return apart from that of 1944, when it was badly placed in the sky and most astronomers had other things on their minds.

In 1682 Edmond Halley, later to become Britain's second Astronomer Royal, observed a bright comet. He was not the first to see it, but he made careful measurements of it, and came to a remarkable conclusion. Up to that time it had been believed that comets moved in straight lines, passing the Sun only once and then departing, never to return; but Halley realized that the 1682 comet moved in the same way as comets previously seen in 1607 and in 1531. Could they be one and the same? Halley thought so – and he was right. He calculated that if the comet had a period of 76 years it would return once more in 1758. By then Halley was dead, but the comet appeared on schedule, and it was only right that it should be called by Halley's name. Since then it has come back in 1835, 1910 and 1986, though on the last occasion it was badly placed and never became brilliant.

Records show that the returns have been regular, and we can trace the comet back well before the time of Christ. It has not always been popular; when it shone forth in 1066, some months before the Battle of Hastings, the Saxon court regarded it as an evil omen, and in 1456 the then Pope, Calixtus III, preached against it as an agent of the Devil. Certainly comets used to be regarded as unlucky; remember the lines in Shakespeare's *Julius Caesar* – 'When beggars die, there are no comets seen; the heavens themselves blaze forth the death of princes.' The fear was partly superstitious, but partly because of the possibility that a direct hit from a comet would mean the end of the world. Ideas of this sort are not quite dead even yet; at the 1986 return of Halley's Comet, several well-meaning organizations loudly proclaimed that the hour of doom was nigh.

In 1986 a whole armada of space-craft was sent to rendezvous with Halley's Comet; there were two Russian probes, two Japanese probes and one European. The European vehicle, Giotto, was built in Britain, and went right inside the comet's head, sending back close-range pictures of a peanut-shaped nucleus made mainly of ices; it was dark, but with dust-jets issuing from localized areas on the sunlit side. The diameter was no more than 14.5 km (9 miles), and it would take 60 000 million Halley's Comets to equal the mass of the Earth; it loses about 300 million tonnes of material at each perihelion passage, so that it cannot last indefinitely. It will be back in 2061, though once again it will be badly placed and will not become brilliant (according to the *Anglo-Saxon Chronicle*, it was so bright at the 837 return that it cast strong shadows).

The Giotto mission was the first successful cometary encounter, but there have been others since. In September 2001, the space-probe Deep Space 1 encountered Borrelly's Comet and sent back images of the nucleus from close range. Borrelly's Comet, originally found in 1904, has a period of 6.9 years; the nucleus is 8 km long and 4 km wide (5 miles by 2^1/$_2$ miles), with a rough surface and deep fractures.

One very interesting experiment was carried out in the early days of the twenty-first century. NASA's Deep Impact space-craft was launched on 12 January 2005 and aimed at a well-known periodical comet, Tempel 1. After a trek of 174 days it neared the comet, and on 3 July separated into two parts, an impactor and a fly-by. The copper impactor, weighing 350 kg (770 lb), hit the comet at a speed of 37 000 kmh (23 000 mph); the fly-by 'dodged' and took pictures of the whole event. Earth-based observers recorded a bright flash; a crater was produced, and a cloud of débris ejected, containing more dust and less ice than expected, but about 250 000 tonnes of water were sent out. It seemed that the comet was about 75 per cent empty space, and the outer layers were compared to the make-up of a snow-bank. Quite unfazed, the comet continued its normal journey around the Sun. There was never any danger of disrupting it, and the impact provided important new information about cometary interiors.

There have been spectacular comets in past years. Chéseaux' Comet of 1744 had at least six tails; the Great Comet of 1811 was visible in broad daylight, as was the Great Comet of 1843, which incidentally sparked off a major end-of-the-world panic in America. Donati's Comet of 1858 is said to have been the most beautiful on record, with a long, straight gas-tail and a scimitar-like dust-tail. The 1882 comet cast shadows, and there was also the Daylight Comet of 1910, which appeared a few months before Halley's. Two bright comets of recent times have been Hyakutake, in 1996, and Hale-Bopp, which became really bright in the spring of 1997 and certainly qualified as a 'great comet'. It remained a naked-eye object for many weeks – but it will not return for some 4000 years. McNaught's Comet of 2007 was brilliant if seen from southern latitudes, but not from the north.

Comet-hunting is a favourite amateur pastime, and although it is laborious it is also rewarding. The procedure is simply to scan the sky with powerful binoculars or a wide-field telescope, noting anything which is unfamiliar. The one essential is an encyclopaedic knowledge of the sky; for example, George Alcock, one of the best modern comet-hunters, knew the positions and magnitudes of over 30 000 stars by heart.

There is no doubt at all of the link between comets and meteors, and one celebrated lost comet, Biela's, has provided us with cast-iron evidence. It used to have a period of 6¾ years, and had been known to reach the third magnitude. When it came back in 1845, it astounded astronomers by breaking in two. The twins returned in 1852, were missed in 1859 because they were so badly placed, and were expected once more in 1866 – but they failed to appear, and have never been seen since. However, in 1872 a shower of meteors was seen to come from the position in the sky where the comet ought to have been, and there is no doubt that these meteors represented the funeral pyre of Biela's Comet.

As we have seen, a meteor is a piece of cometary débris, and when we pass through the 'dusty' trail left by a comet we see a shower of shooting-stars. This happens many times in every year. The meteors of a shower appear to radiate from a definite point in the sky, because they are travelling through space in parallel paths; a good way to demonstrate this is to stand on a bridge overlooking a motorway, and see how the parallel lanes appear to radiate from a point in the distance. Each shower is named after the constellation which contains the radiant. Thus the August meteors come from Perseus, and are known as the Perseids (see Figure 11.3); the November Leonids come from Leo, and so on.

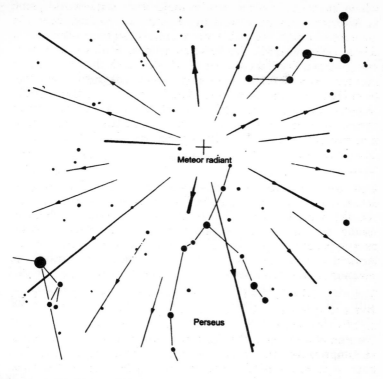

Meteor radiant

Perseus

Figure 11.3 Meteor radiant in Perseus

Not all showers are equally rich. They are classified according
to the ZHR or Zenithal Hourly Rate, which is defined as being
the number of naked-eye shower meteors which would be
expected to be seen by an observer under ideal conditions, with
the radiant at the zenith or overhead point. In practice these
conditions are never attained, so that the actual ZHR is lower
than the theoretical limit. A list of the principal meteor showers
is given in Appendix 4. (Note that the Quadrantid radiant lies
in the constellation of Boötes, the Herdsman; it is the site of an
old constellation, Quadrans (the Quadrant) which has been
deleted from modern maps.)

The Perseids are the most spectacular of the annual showers,
because they have been spread all around the orbit of the parent
comet, Swift–Tuttle. Stare up into a clear, dark sky at any time
during the first part of August, and you will be unlucky not to

see several Perseids. The November Leonids are quite different; they are still 'bunched up', and are rich only when the Earth ploughs through the thickest part of the swarm, as in 1799, 1833, 1866 and 1966. On these occasions it is said that meteors 'rain down from the sky like snowflakes', though unfortunately the 1966 display took place during daylight in Europe, and was seen only from the other side of the world. The parent comet, Tempel–Tuttle returned to perihelion in 1999 and there were spectacular Leonid displays in 2000 and 2001, though no more can be expected until the comet comes back once more.

Amateurs carry out valuable work in meteor observing. It is true that radar methods have to some extent taken over, because a meteor trail reflects radar pulses, but visual observations are still useful. The best way to plot the track of a meteor is to hold up a ruler against the path of the shooting-star, so that the beginning and end points can be found with reference to the starry background; again a good knowledge of the sky is needed. Note also the exact time, duration of visibility, brightness, colour (if any) and any other special characteristics. If the same meteor is observed by observers at different sites, it is possible to work out the height by the method of triangulation.

Meteor photography is fascinating; simply point the camera in the appropriate direction, open the shutter and give as long a time exposure as you can without getting dewed up, so that you will 'net' any meteor which happens to flash through the field of view. Of course, not all meteors are members of known showers; there are also sporadic meteors which may appear from any direction at any moment, and all in all it is estimated that 75 million naked-eye meteors enter the Earth's air every 24 hours. Under normal conditions a dedicated observer may expect to see around ten meteors per hour, though during a major shower the number is naturally much higher. And now and then a really magnificent meteor appears; I have seen four which became decidedly brighter than the full moon.

Meteorites are larger bodies, which survive the complete drop to the ground without being burned away. Note that a meteorite is not simply a large meteor, and in fact there is no connection at all. Meteorites come from the asteroid belt, and there is no true distinction between a large meteorite and a small asteroid; it is a question of terminology, though the term 'meteorite' is conventionally limited to objects which have landed on Earth.

Many thousands of meteorites have been collected, and some of these have been observed during their descent. Such was the small meteorite which shot across the English sky on Christmas Eve 1965; it broke up before impact, and distributed fragments around the Leicestershire village of Barwell in England. There are no reliable reports of any human death by meteorite fall, though it is true that one or two people have had narrow escapes.

Meteorites are of two main types: stones (aerolites) and irons (siderites). They are not always easy to identify, but geologists can recognize them, and if an iron meteorite is cut and then etched with acid it will show characteristic features known as Widmanstätten patterns, not found in terrestrial minerals. Some stony meteorites, known as chondites, contain small spherical particles (chondrules) made up of fragments of minerals; with other meteorites, chondrules are absent. It has been suggested that some meteorites have been knocked off the Moon or even Mars.

The largest known meteorite is still lying where it fell, in prehistoric times, near Grootfontein in Southern Africa; to move it would be difficult, because its weight seems to be over 60 tonnes. The largest meteorite on display is the Ahnighito or 'Tent', which was found in Greenland by the explorer Robert Peary and taken to the Hayden Planetarium in New York, where you can now see it.

Go to Arizona, not far from the town of Winslow, and some way off Highway 99 you will come to a huge crater, the best part of 1.6 km (1 mile) wide, which was certainly made by a meteorite impact over 50 000 years ago. It has become a well-known tourist attraction, and is worth visiting; the Swedish scientist Svante Arrhenius once described it as 'the most interesting place on earth'. Wolf Creek Crater, in Western Australia, is not so large and is much less accessible, but it is well-formed, and there is no doubt about its origin. Other impact craters are found in the Northern Territory of Australia, Waqar in Arabia, Oesel in Estonia, and here and there in the United States.

There is considerable doubt about the nature of the missile which hit the Tunguska region of Siberia on 30 June 1908. Pine trees were blown flat over a wide area, and the noise was heard hundreds of kilometres away. It was lucky that nobody lived there; if the object had hit a city, the death-toll would have been colossal. No crater was produced, and it is unclear whether the impactor has to be classed as a meteorite, a tiny asteroid, or a piece of cometary débris. A second Siberian fall occurred on 12 February 1947, in the Vladivostok area, but there is no mystery about this, as over 100 crater-pits were found and many meteorite fragments collected.

I must also refer briefly to tektites, which are small, glassy objects found in certain restricted areas, such as Australasia and the Ivory Coast. They are aerodynamically shaped, and give every indication of having been heated twice. They were once believed to be meteoritic, but it now seems that they are of terrestrial origin, and were shot out from volcanoes.

There is a great deal of thinly spread interplanetary material in the Solar System. This is the cause of the Zodiacal Light, a faint, luminous cone seen rising from the horizon soon after sunset or soon before sunrise and extending along the Zodiac – hence its name. It is not easy to see from light-polluted Britain, but from countries with clearer skies it can be very prominent; it is due to sunlight catching particles scattered along the main plane of the Solar System. The Gegenschein or Counterglow, a very dim patch of light exactly opposite to the Sun in the sky, is of similar origin; it is very elusive, and I have seen it only once in my life.

All in all, the Solar System is a fascinating place, and to us it is of supreme importance. Yet we have to concede that it is a very minor feature of the Galaxy, and it is now time to turn our attention to realms much further afield – far away in the depths of space.

12

the stars

In this chapter you will learn:
- about the stars, and what they are
- how we measure the stars' distances
- how stars are divided up into constellations.

No ordinary telescope will show a star as anything other than a point of light; in fact, the smaller a star looks, the better you are seeing it! This is not because the stars are small, but because they are so remote. Measuring their distances is no easy matter, and it defeated the skills even of Sir William Herschel. Success came finally in 1838, due to the careful work carried out by F. W. Bessel in Germany.

Bessel decided to use the method of triangulation. Essentially, this is the same as the method used by a surveyor to work out the distance of some inaccessible object, such as a mountain-top. In Figure 12.1, the observer (O) measures the angle of the mountain as seen from two points, A and B, which are separated by a known distance. The angles BAM and MAB can be found, and hence the angle BMA. Half this angle is known as the parallax. The distance MO can then be found by straightforward mathematics, and this of course is what the surveyor needs.

To measure the distance of a star we require a much longer baseline, and Bessel used the diameter of the Earth's orbit. He selected a dim star, 61 Cygni in the constellation of the Swan, which he had reason to believe must be relatively nearby, and then measured its position against the background of more remote stars, first in January and then in July. During the six-months' interval the Earth moved from one side of its orbit to the other, giving Bessel a baseline of 300 million km (186 million miles). The parallax shift of 61 Cygni was very small, less than one-third of a second of arc, but it was definite enough, and from it Bessel was able to show that the distance of 61 Cygni is just over 11 light-years, corresponding to roughly 103 million million km (64 million million miles). Not many stars are as near as that; the closest of all, the dim red dwarf Proxima Centauri in the southern sky, has a parallax of 0.76 of a second of arc, corresponding to 4.25 light-years.

Bessel's reasons for selecting 61 Cygni were twofold. First, the star has a definite proper motion – 4.1 seconds of arc per year; secondly, it was made up of two components separated by almost 30 seconds of arc, so that a small telescope will show them individually. Since both are feeble and red, the fact that they were so far apart gave another indication that they could not be too far from the Sun – at least by cosmical standards.

At the Cape Observatory in South Africa, Thomas Henderson had been making similar measurements of the brilliant Alpha Centauri, one of the two pointers to the Southern Cross, which also has considerable proper motion (in this case 3.6 seconds of

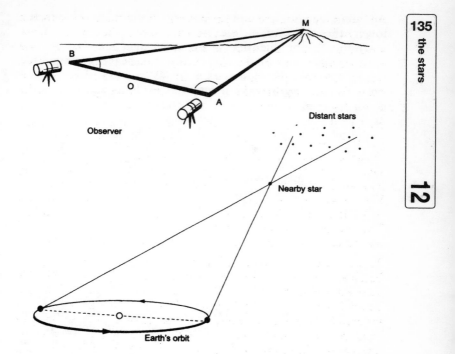

Figure 12.1 Method of parallax

arc annually) and is a wide double. Henderson's work was carried out before Bessel's, and was easier, because Alpha Centauri is closer to us – and is only one-tenth of a light-year further away than Proxima, which is a member of the Alpha Centauri group. However, Henderson was in no hurry to publish his results, and failed to do so before Bessel's announcement in 1838.

The parallax method works well to distances of a few hundreds of light-years, but after that the shifts are swamped by unavoidable errors in observation, so that we have recourse to less direct methods. Most of these involve using spectroscopic analysis to find out how luminous the star really is; we can then compare its apparent magnitude with what is termed its absolute magnitude, and derive the distance. Some stars, known as Cepheid variables, make things easier because they 'give away' their luminosities by the way in which they behave.

Apparent magnitude is a measure of how bright a star looks; the lower the magnitude, the more brilliant the star. Thus magnitude 1 is brighter than 2, 2 brighter than 3, and so on. The most brilliant star in the sky, Sirius, has negative magnitude (−1.5), while the dimmest stars normally visible with the naked eye on a clear night are of magnitude +6. Binoculars can reach down to about magnitude 9, and today electronic equipment used with the world's largest telescopes can penetrate down to magnitude 30. Conventionally, the 21 brightest stars are said to be of the 'first magnitude', ranging from Sirius down to Regulus in Leo, the Lion, whose magnitude is 1.3, but there are 29 more stars between magnitudes 1.5 and 2. The magnitude of the Pole Star is 2.0; the scale is logarithmic, and a star of magnitude 1.0 is exactly 100 times brighter than a star of magnitude 6.0. A list of the first-magnitude stars is given in Appendix 6.

The apparent magnitude of a star has only a limited connection with its real luminosity, because the stars are at very different distances from us. Consider, for example, Sirius (−1.5) and its nearest rival, Canopus, in the southern sky (−0.7). Sirius looks brighter by almost a magnitude − but it is only 8.6 light-years away, and no more than 26 times as powerful as the Sun, while Canopus is over 1000 light-years away and could match 200 000 Suns. (The distances of very remote stars such as Canopus are obviously hard to measure accurately, and different authorities give different values.) As so often in astronomy, appearances can be deceptive.

The absolute magnitude of a star is the apparent magnitude which it would have if we could see it from a standard distance of 32.6 light-years. From this range Sirius would be of magnitude +1.4, and would just qualify as being of the 'first magnitude', while Canopus would blaze down at magnitude −8.5, and would be the most brilliant object in the sky apart from the Sun and the Moon. The absolute magnitude of the Sun is +4.8, so that it would appear only as a dim naked-eye object.

Why 32.6 light-years? Well, if a star showed a parallax of 1 second of arc, its distance would be 3.26 light-years − a unit known as the parsec. In fact no star apart from the Sun is as close as this, but it explains why we use the standard distance of 32.6 light-years, or 10 parsecs, to determine stellar absolute magnitudes.

The holder of the 'speed record' is Barnard's Star, a dim red dwarf which is only 5.8 light-years away and is the closest of all stars apart from the three members of the Alpha Centauri system, but even here the proper motion is only just over 10 seconds of arc per year, so that it takes 190 years for Barnard's Star to crawl across its background by a distance equal to the apparent diameter of the full moon. This is why the constellation patterns do not change appreciably over periods of many lifetimes. Eventually, of course, they will alter; for example, in the famous Plough pattern of Ursa Major two of the members are moving in a direction opposite to that of the other five, so that if we could come back in (say) 50 000 years the shape would be unrecognizable. But as far as we are concerned, it is fair to say that the constellations are permanent.

Yet a constellation has no real significance, and we are dealing with nothing more than line-of-sight effects. The Chinese and the Egyptians had their own constellation patterns, and if we followed one of their systems our sky-maps would look very different, though the stars would be exactly the same. It so happens that we use the Greek patterns, though with Latin names. Ptolemy, last of the great astronomers of ancient times, listed 48 groups, all of which are still in use though in many cases with altered outlines.

New constellations have been added since Ptolemy's time, notably in the far south of the sky which was inaccessible from Egypt – where Ptolemy is believed to have spent all his life. Moreover, there was a period when astronomers felt bound to add new constellations, stealing stars from accepted groups, so that we have decidedly modern names such as the Telescope (Telescopium) and Antlia (the Air-pump). The eventual sky-map was – and still is – a muddle; Sir John Herschel, son of William, once commented that the constellations seemed to have been designed so as to cause the maximum possible inconvenience. Then, in the 1930s, the International Astronomical Union lost patience and revised the outlines, rejecting some of the most insignificant groups with barbarous names (such as Sceptrum Brandenburgicum and Globus Aerostaticus), and modifying others; the largest of all Ptolemy's constellations, Argo Navis (the Ship Argo) was chopped up into a keel, sails and a poop. Today we recognize 88 constellations, diverse in size and importance.

In 1603 a German amateur astronomer, Johann Bayer, compiled a new star atlas and introduced the system of allotting Greek letters to the stars. Thus the brightest star in a constellation was lettered Alpha, the second brightest Beta, and so on down to Omega, the last letter of the Greek alphabet. Sirius in Canis Major (the Great Dog) therefore becomes Alpha Canis Majoris.

In many cases the letters are out of sequence; for example, in Sagittarius (the Archer) the brightest stars are Epsilon and Sigma, with Alpha and Beta very much 'also rans'. Still, in general the system is a good one, and it has stood the test of time. Years later John Flamsteed, the first Astronomer Royal, gave the stars numbers in order of right ascension, so that for example Sirius is 9 Canis Majoris, and Polaris is both Alpha Ursae Minoris (Alpha of the Little Bear) and 1 Ursae Minoris.

Since the Greek letters will be so often used in the following pages, it may be wise to list them here. They are:

α	Alpha	ν	Nu
β	Beta	ξ	Xi
γ	Gamma	ο	Omicron
δ	Delta	π	Pi
ε	Epsilon	ρ	Rho
ζ	Zeta	σ	Sigma
η	Eta	τ	Tau
θ	Theta	υ	Upsilon
ι	Iota	φ	Phi
κ	Kappa	ξ	Chi
λ	Lambda	ψ	Psi
μ	Mu	ω	Omega

Individual star names are generally used only for the 21 stars of the first magnitude, plus a few special cases such as the famous double Mizar (Zeta Ursae Majoris) and the red variable Mira (Omicron Ceti). Most of the names are Arabic, and date back around a thousand years, though a few are older; Sirius is from the Greek word meaning 'scorching', and should be pronounced with a long ī, though most people refer to it as 'Sirrius'. There are cases of stars with several proper names, and there are various spellings and pronunciations; the name of Alpha Orionis, the red supergiant in the Hunter's shoulder, may be spelled Betelgeux, Betelgeuze or Betelgeuse. Nobody has ever been quite sure how to pronounce it; it has even been referred to as 'Beetlejuice'!

The names of fainter stars have fallen into disuse, though they are still met with occasionally. (Incidentally, beware of unscrupulous organizations which claim to be able to name stars upon payment of sums of money. These names are quite meaningless and unofficial, and, to be frank, schemes of this sort can only be described as confidence tricks. Have nothing to do with them.)

Stars are of many kinds. There are doubles, most of which are physically associated or are binary systems; there are variables, which change in light over short periods either regularly or erratically; there are giants and dwarfs; there are clusters of stars, some 'loose' and some symmetrical or globular; and there are the huge gas-and-dust patches known as nebulae. And, of course, far beyond our Milky Way system there are the outer galaxies, so remote that in most cases their light takes millions of years to reach us. The stellar sky is full of interest, and so let us now go on a somewhat brief tour. After all, the stars become so much more fascinating when you know which is which.

13

patterns of stars

Finding your way around the sky is not nearly so difficult as might be thought. There are only a few thousand stars visible with the naked eye, and the brighter ones stand out at once; when you have identified a constellation you will always be able to recognize it again, because it will not alter (unless a planet moves into it). I well remember that when I set out to learn the patterns, at the advanced age of eight, I made a pious resolve to identify one new constellation every clear night, and after a few weeks I was fairly confident that I could recognize all the main groups.

I am not here setting out to provide a complete guide to the night sky; other books do that. All I hope to provide is some basic information, using the Latin names of the constellations (a full list, with the English equivalents, is given in Appendix 5) and adding star magnitudes in brackets. So let us begin by observing from the latitude of London, which is approximately 51 degrees north. The method I have always found convenient is to select a few groups which are quite unmistakable, and use them as guides to the rest; much the most useful of these guides are Ursa Major (the Great Bear) and Orion.

Winter evenings (northern hemisphere)

This is a good time to commence operations, because both Ursa Major and Orion are on view; Ursa Major stands 'tail down' in the north-east, while Orion is high in the south (see Figure 13.1). Ursa Major never sets over any part of the British Isles, though Orion is out of view for some months during the summer.

The seven chief stars of Ursa Major make up the pattern known familiarly as the Plough or, in America, the Big Dipper. All except one are of around the second magnitude, and though they are not outstandingly bright they are very easy to identify because they lie in an otherwise rather barren area. Look at Zeta or Mizar (2.1), the second star in the Bear's tail, and find the fourth-magnitude companion, Alcor; a telescope will show that Mizar is itself double, with one component decidedly brighter than the other. Between Alcor and the Mizar pair there is a fainter star, which is much more remote from us and happens to lie more or less in the same line of sight.

The two end stars of the pattern, Alpha or Dubhe (magnitude 1.8) and Beta or Merak (2.4), are known as the Pointers because they show the way to Polaris, the Pole Star (2.0) in the Little

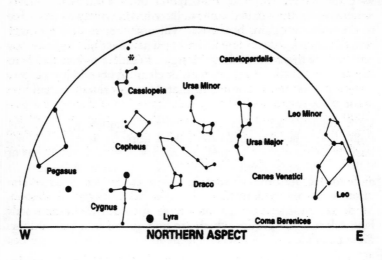

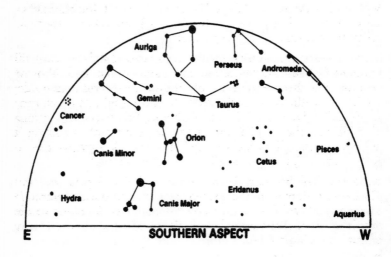

Figure 13.1 Northern winter sky chart

Bear (Ursa Minor). Even with the naked eye you can see that Dubhe and Merak are not alike; Dubhe is orange, while Merak is pure white, showing that Dubhe has the lower surface temperature, though it is considerably further away from us and is the more luminous of the two. The other stars of the Plough are Epsilon or Alioth (1.8), Eta or Alkaid (1.9), Zeta, Gamma or Phad (2.4) and Delta or Megrez (3.3). Though it has been given the fourth letter of the Greek alphabet, Megrez is a magnitude fainter than the other members of the Plough. There have been suggestions that it may have faded during the past 2000 years, but this does not seem likely because it is not the kind of star which would be expected to show secular variation over a period of a few centuries. It is 65 light-years away, and 17 times as luminous as the Sun.

Follow the line of the Pointers, and locate Polaris. Ursa Minor bears a slight resemblance to a dim and distorted Plough, curving down towards Mizar and Alkaid, but it contains only one other fairly bright star, Beta or Kocab (2.1); like Dubhe, Kocab is decidedly orange.

Draco (the Dragon) is a long, sprawling constellation in this part of the sky. It contains no really bright star, but there is one which is worth locating. Almost between Alkaid and Kocab you will find Thuban or Alpha Draconis (3.6). It looks, and is, unremarkable; but it used to be the north pole star at the time when the Pyramids were being built.

If you continue the line of the Pointers beyond Polaris, you will come to Cassiopeia, whose five main stars make up a W or M pattern. The three leaders are of around the second magnitude; Alpha or Shedir is decidedly reddish, and though its magnitude is officially given as 2.2 it is probably slightly variable. Gamma Cassiopeiae, the middle star of the W, is certainly variable; it suffers occasional outbursts, and been known to rise to magnitude 1.6, though usually it is rather below 2.

More or less between Ursa Minor and Cassiopeia we find Cepheus, notable for the presence of two important variable stars, one of which – Delta Cephei – has given its name to the whole class of short-period variables. I will have more to say about it in Chapter 14.

By late January Leo, the Lion, is coming into view during evenings; by the end of the winter it will be high up after dark. To find it, use the Pointers in the direction away from Polaris; the alignment is not perfect, but it is good enough. In the west the Square of Pegasus, the main autumn constellation, is setting; two of the stars in the W of Cassiopeia can be used to locate it.

Now turn to Orion, high in the south (see Figures 13.1 and 13.2). Here we have an unmistakable pattern, with two brilliant stars, Alpha or Betelgeux (variable, but usually about 0.6) and Beta or Rigel (0.1). The others stars in the main pattern are Gamma or Bellatrix (1.6), Kappa or Saiph (2.1) and the three members of the Hunter's Belt, Epsilon or Alnilam (1.7), Zeta or Alnitak (1.8) and Delta or Mintaka (2.2). All these are hot and bluish-white; Mintaka lies practically on the celestial equator. Rigel – which should logically be lettered Alpha, rather than Beta – is a veritable cosmic searchlight, at least 60 000 times more powerful than the Sun and around 900 light-years away.

Below the Belt you will see a misty patch. This is the Great Nebula, No. 42 in a famous catalogue compiled in 1781 by the French astronomer Charles Messier. It is a stellar birthplace, well over 1000 light-years away, shining because it is being lit up by a multiple star, Theta Orionis, on its Earth-turned side.

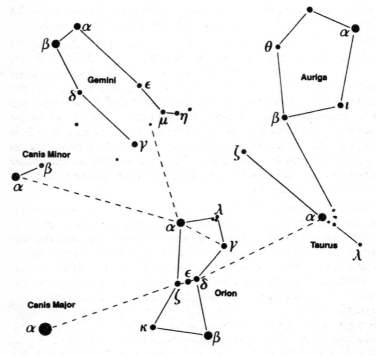

Figure 13.2 Orion area

Orion's Belt points downward to Sirius in Canis Major, which is much the most brilliant star in the sky (−1.5). It is pure white, but because it is always rather low down as seen from Britain it seems to flash various colours of the rainbow. Star twinkling has nothing directly to do with the stars themselves; it is due entirely to the Earth's unsteady atmosphere, which, so to speak, 'shakes' the light around. The greater the altitude of the star above the horizon, the less it will twinkle. Canis Major is a large constellation with several more bright stars, but it is completely dominated by Sirius.

Next, use Orion to find Procyon (0.4), Alpha of Canis Minor, the Little Dog. Also in Orion's retinue we find the Twins, Castor and Pollux, in the constellation of Gemini. The two are not true neighbours; Pollux is 36 light-years away from us, Castor 45, and Castor (1.6) is appreciably fainter than Pollux (1.1). Pollux is orange, while Castor is white − and is, incidentally, a fine binary.

Gemini is crossed by the Milky Way, and is marked by lines of stars extending from Castor and Pollux in the general direction of Betelgeux. Gamma or Alhena (1.9) is prominent, and at the 'foot' of the Twins we find a lovely open star-cluster, M.35.

Upward, the three stars of Orion's Belt show the way to Aldebaran or Alpha Tauri (0.8), the 'Eye of the Bull'. It is not nearly so luminous as Betelgeux, but it is of about the same colour. Extending in a V-formation away from Aldebaran is the open cluster of the Hyades, so spread-out that it is better seen with binoculars than in a telescope. Further away from Orion lies the open cluster of the Pleiades or Seven Sisters, which is without doubt the most celebrated cluster of its type, and is prominent with the naked eye. Taurus itself has no particular shape, but the presence of Aldebaran and the two open clusters makes it unmistakable.

Capella, in Auriga (the Charioteer) is almost overhead during winter evenings; its magnitude is 0.1, and, like the Sun, it is yellowish in colour, though it is far more luminous than the Sun and is actually a very close binary. Auriga is marked by a quadrilateral of stars, one of which, Al Nath, is bright (1.6). Al Nath used to be included in Auriga, as Gamma Aurigae, but for some unknown reason it has been given a free transfer to Taurus, and is now known officially as Beta Tauri. Adjoining Auriga is the rich constellation of Perseus, led by Alpha or Mirphak (1.8). The most famous object here is Beta or Algol, which is normally of the second magnitude, but which fades down to below magnitude 3 every two and a half days – because it is a binary system, and drops in brightness when the fainter component passes in front of the primary and eclipses it.

Between Perseus and the Square of Pegasus we find Andromeda, marked by a line of brightish stars, which is at its best during evenings in autumn. In the south-west lies the rather barren constellation of Cetus (the Whale), and below Andromeda is Aries (the Ram), which is always regarded as the first constellation of the Zodiac even though the vernal equinox has now shifted into Pisces (the Fishes), which is very obscure indeed.

Spring evenings (northern hemisphere)

By late evenings in April Orion has almost disappeared in the west, though parts of its retinue remain prominent. Ursa Major is almost at the zenith, with Capella in the north-west (see Figure 13.3).

Follow round the 'tail' of the Great Bear (Alioth, Mizar, Alkaid) and you will come to a brilliant orange star, Arcturus. It is slightly above zero magnitude (–0.04), and is the brightest star visible from Britain apart from Sirius. It is the leader of Boötes, the Herdsman, which is well north of the celestial equator even though most of it is not circumpolar from the latitude of London. It has a distinctive pattern resembling a Y, but one of the branches of the Y belongs not to Boötes but to the small constellation of Corona Borealis (the Northern Crown), which is one of the few groups to bear some resemblance to the object after which it is named. The little semi-circle of stars is easy to identify, though only one, Alphekka or Alpha Coronae, is of the second magnitude.

Follow the Ursa Major–Arcturus line further, curving it somewhat, and you will come to Spica (1.0) in Virgo, the Virgin. Spica is a white star, more than 2000 times as luminous as the Sun and 250 light-years away; it is actually a very close binary. The rest of Virgo consists of a well-marked Y, covering a large area. The star at the base of the Y, Gamma Virginis or Arich (2.7) is a splendid double with almost equal components. Inside the 'bowl' of Virgo are many faint galaxies, and between Virgo and Ursa Major is Coma Berenices (Berenice's Hair), a loose, extended cluster which looks like a patch of soft radiance in the sky.

As we have seen, Leo (the Lion) can be located by using the Pointers in Ursa Major in the direction away from Polaris. The

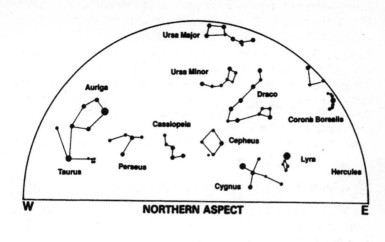

NORTHERN ASPECT

W   E

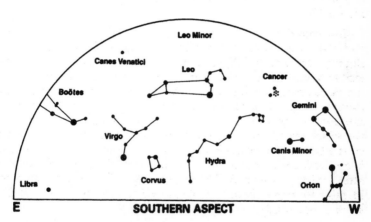

SOUTHERN ASPECT

E   W

Figure 13.3 Northern spring sky chart

main part of Leo is marked by the 'Sickle', shaped rather like the mirror image of a question-mark; Regulus (1.3) is the brightest member, while Gamma Leonis or Algieba (2.0) is a fine telescopic binary (see Figure 13.4). Some way from the Sickle is a triangle of stars: Beta or Denebola (2.1), Delta (2.6) and Theta (3.3). Many of the old astronomers ranked Denebola as being of the first magnitude, equal to Regulus. It is now much fainter, though whether any real change has taken place is doubtful, and more probably we are dealing with mistakes in translation or interpretation.

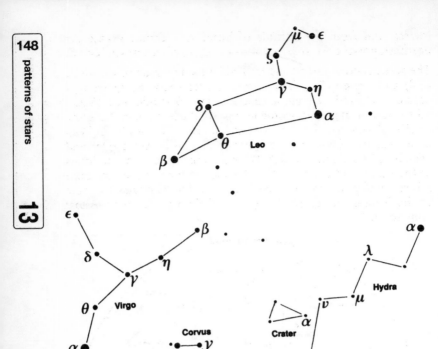

Figure 13.4 Leo area

Next, locate Alphard or Alpha Hydrae (2.0), in the Watersnake. It is clearly reddish, and is known as the 'Solitary One' because there are no other bright stars anywhere near it. The best way to find it is to use the Twins, Castor and Pollux, as pointers. Hydra is the largest constellation in the sky, winding from near Gemini right down to the south of Virgo, but it contains little of immediate interest. Adjoining it are two small constellations, Crater (the Cup) and Corvus (the Crow). Crater is very obscure, but Corvus is conspicuous enough; it lies near Spica, and its four main stars, all between magnitudes 2½ and 3, are easy to identify.

Summer evenings (northern hemisphere)

Orion has now disappeared; Ursa Major is in the north-west, while Leo and Virgo are low in the west and Pegasus is starting to come into view in the east (see Figure 13.5). Capella is at its lowest, and from the latitude of London it almost grazes the northern horizon, so that any mist or light pollution will hide it.

The scene is dominated by three brilliant stars: Vega (0.0), Altair (0.8) and Deneb (1.2). Years ago, in a television programme, I casually referred to them as the 'Summer Triangle' (see Figure 13.6), and the name has come into general use, though it is quite unofficial and does not apply to the southern hemisphere. Vega is in Lyra (the Lyre or Harp), Altair in Aquila (the Eagle) and Deneb in Cygnus (the Swan). Deneb is much the most luminous of the three, and is about 70 000 times as powerful as the Sun; it is so remote that we now see it as it used to be during the time of the Roman occupation. Vega is 26 light-years away, Altair only about 17.

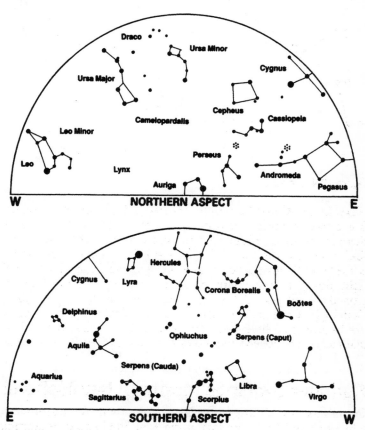

Figure 13.5 Northern summer sky chart

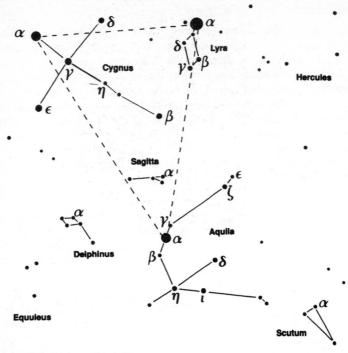

Figure 13.6 Summer Triangle area

Vega, almost overhead during summer evenings, is of a beautiful blue colour, so that with binoculars it is a splendid sight. Lyra is a small constellation, but contains a surprising number of interesting objects; for example, the double-double star Epsilon Lyrae, the eclipsing variable Beta Lyrae, and the most famous of all planetary nebulae, M.57 (the Ring), which is an easy telescopic object. Like all planetaries it is not a true nebula, but merely a very old star which has blown away its outer layers.

Altair (Alpha Aquilae) is lower down, and is the only member of the Summer Triangle which is not circumpolar from London. It is particularly easy to identify because it is flanked to either side by a fainter star, Gamma (2.7), which is orange, and the white Beta (3.7). Aquila itself is distinctive, with a pattern which really does conjure up the vague impression of a bird in flight. At its southern end the Milky Way is very rich, and here we have the small constellation of Scutum (the Shield), containing the cluster M.16, often nicknamed the Wild Duck.

Cygnus is known as the Northern Cross, for obvious reasons. The central star of the X is Gamma (2.2). One member of the pattern, Albireo or Beta Cygni (3.1), is fainter than the rest and further away from the centre so that it spoils the symmetry, but to make up for this it proves to be a superb coloured double, with a golden-yellow primary and a vivid blue companion. Also in Cygnus there are dark rifts in the Milky Way, due to masses of opaque material blotting out the light of stars beyond. There are several small constellations in the general region, one of which, Delphinus (the Dolphin), looks almost like a very open cluster; newcomers to the night sky have often mistaken it for the Pleiades.

The region enclosed by lines joining Vega, Altair and Arcturus is occupied by three large but dim and ill-formed constellations: Hercules, Ophiuchus (the Serpent-bearer) and Serpens (the Serpent). The only bright star is Rasalhague or Alpha Opriuchi (2.1). Frankly, this is a rather confusing area, and there are not many objects of immediate note other than the globular cluster M.13 in Hercules, which is just visible with the naked eye on a clear night. Rasalgethi or Alpha Herculis, near Rasalhague, is a red variable ranging between magnitudes 3 and 4.

Very low in the south are two magnificent Zodiacal constellations, Scorpius (the Scorpion) – the constellation is often referred to as Scorpio, but Scorpius is the correct name – and Sagittarius (the Archer), parts of which never rise at all over London. Antares (1.0) is the leader of Scorpius; it is fiery red – its name means 'the Rival of Ares' (Mars) – and like Altair it has a fainter star to either side of it. There are many interesting objects in the Scorpion, such as the two naked-eye open clusters M.6 (the Butterfly) and M.7 (Ptolemy's cluster). There is no mistaking the line of bright stars which characterizes Scorpius, but from Britain's latitude it is never really well seen. Sagittarius has no obvious shape, though it is often nicknamed the Teapot; its brightest star, which is easily visible from Britain, is Sigma or Nunki (2.0). The Milky Way is exceptionally rich in Sagittarius, and there are glorious 'star clouds' which block Britain's view of that mysterious region, the centre of the Galaxy.

Between Scorpius and Virgo lies the obscure Zodiacal constellation of Libra (the Scales or Balance). Its brightest star, Beta, is said to be the only naked-eye star with a greenish tinge, though most people will certainly call it white. Note that Ophiuchus intrudes into the Zodiac between Scorpius and Sagittarius, so that planets can pass through it even though it is not officially classed as a Zodiacal constellation.

Autumn evenings (northern hemisphere)

It is true to say that the evening sky in autumn is less brilliant than at any other time in the year (see Figure 13.7). Scorpius and Sagittarius have disappeared; Orion has not yet come into view, and Ursa Major is at its lowest in the north, though it remains well above the horizon. The Summer Triangle is still on view, now west of the zenith, and Capella is rising in the northeast. Capella and Vega are on opposite sides of the Pole Star, and about the same distance from it, so that when Vega is high up Capella is low down, and vice versa.

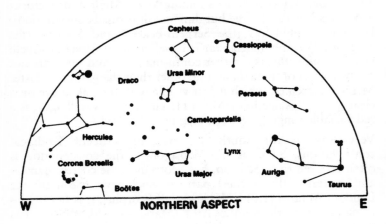

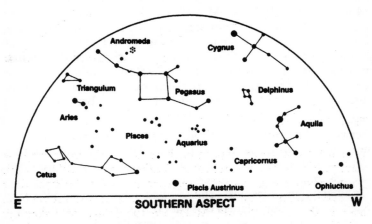

Figure 13.7 Northern autumn sky chart

The main autumn constellation is Pegasus, named after the mythological flying horse (see Figure 13.8). The four main stars make up a square, which is distinctive enough even though maps tend to make it look smaller and brighter than it really is. The members of the Square are Alpheratz (2.1), Beta Pegasi or Scheat (variable 2.3 to 2.8), Alpha Pegasi (2.5) and Gamma Pegasi (2.8). Alpheratz was once known as Delta Pegasi, but another dubious transfer has given it to Andromeda, as Alpha Andromedae. The only other bright star in Pegasus is Epsilon (2.4), well to the west of the Square.

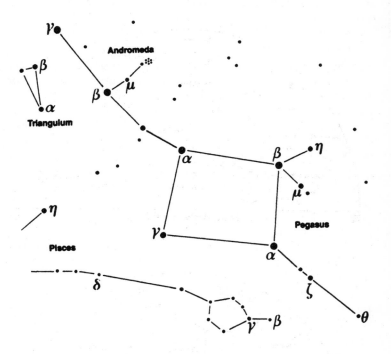

Figure 13.8 Pegasus area

Pegasus is not a rich area, and there are not many naked-eye stars inside the Square. Scheat is clearly orange in colour, while the other three members of the Square are white. Andromeda is marked by a line of fairly bright stars: the most celebrated object here is the Great Spiral, M.31, on the fringe of naked-eye visibility.

Pegasus can be used to find Fomalhaut in Piscis Australis, the Southern Fish, which is the most southerly of the first-magnitude stars to be visible from Britain. It is a relatively close neighbour, at only 22 light-years; it is white, and 13 times as luminous as the Sun.

Scheat and Alpha Pegasi, in the Square, point downward to Fomalhaut. Avoid confusion with Diphda or Beta Ceti (2.0), which is roughly indicated by a line passed from Alpheratz through Gamma Pegasi and extended; Diphda is higher up than Fomalhaut, and almost a magnitude fainter. Cetus (the Whale) is large but dim; its most famous object is Mira, the long-period variable, which has been known to reach magnitude 1.6 at maximum, but which is visible with the naked eye for only a few weeks in every year.

Finally, look for three large, faint Zodiacal constellations. Pisces (the Fishes) consists of a long line of faint stars below Pegasus; Aquarius (the Water-bearer) and Capricornus (the Sea-goat) lie more or less between Pegasus and Fomalhaut. Alpha Capricorni is a naked-eye double. There is little of interest here, but near Beta Aquarii (2.9) we find the magnificent globular cluster M.2, easily visible with binoculars.

This description of the evening sky throughout the year is very incomplete, and many of the smaller constellations have not been mentioned at all, but it should serve as a starting-point, and once you have identified the main groups you will be ready to turn to a more detailed star-map. So let us now turn to the southern hemisphere, and station ourselves at, say, the latitude of Johannesburg (26 degrees south). For most purposes the following notes can be taken to apply to most of South Africa and Australia; some modification is needed for New Zealand – for example Capella never rises over part of South Island, though it can be glimpsed from most of the rest of the country.

One immediate problem is that there is no bright south polar star. The nearest candidate is the obscure Sigma Octantis (5.5), and the whole polar region is depressingly barren. Orion is invaluable when it can be seen, but of course it is out of sight during the winter. Ursa Major is useless. It can attain a respectable altitude from Darwin, and parts of it can be seen over most parts of South Africa and Australia, but it skirts the horizon and is never visible for long; from New Zealand it is to all intents and purposes lost altogether.

The most famous group of the far south is of course Crux Australis, which is actually the smallest constellation in the entire sky, but is crammed with interesting objects (see Figure 13.9). In shape it is more like a kite than a cross, because it lacks any central star, and in any case the symmetry is spoiled by the fact that one member of the pattern is much fainter than the other three. The main stars are Acrux or Alpha Crucis (0.8), Beta (1.2), Gamma (1.6) and Delta (2.8). Three of these are hot and white, but the fourth, Gamma Crucis, is a red giant; binoculars bring out the colour beautifully. Alpha Crucis is a wide double, separable with almost any telescope and with a third star in the field. Note also the lovely cluster around Kappa Crucis, nicknamed the Jewel Box, and the dark nebula which is known as the Coal Sack.

Crux is almost surrounded by Centaurus (the Centaur), and until 1679 was included in it. The brilliant stars Alpha Centauri (–0.3) and Beta (0.6) show the way to the Cross, and are known as the Pointers. Once again appearances are misleading; Alpha Centauri is the nearest bright star beyond the Sun, only a little over 4 light-years away, while Beta is a real giant, over 10 000 times as luminous as the Sun and about 450 light-years from us.

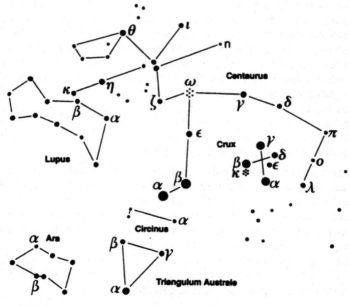

Figure 13.9 Crux area

Although Alpha Centauri is the brightest star in the sky apart from Sirius and Canopus, it has no universally accepted proper name. Some air navigators call it Rigel Kent, and others refer to it as Toliman, but astronomers prefer to call it simply Alpha Centauri. Like Alpha Crucis, it is a wide, easy double. The nearest star to the Solar System, Proxima, is a member of the same group, but is very faint indeed; whether it is genuinely associated with Alpha is by no means certain. Beta Centauri is often called Agena, but has an alternative proper name, Hadar.

Achernar, in Eridanus (the River) lies on the opposite side of the pole; its magnitude is 0.5, and it is almost 800 times as powerful as the Sun. One of the best ways to locate the polar region is to look directly between Crux and Achernar, but there are no bright stars there, and the whole region appears blank when there is the slightest mist or fog. It contains various small, dim and shapeless constellations with modern-sounding names. In Octans (the Octant) Bayer's system has really broken down; the brightest star in the constellation is Nu (3.8).

Summer evenings (southern hemisphere)

At the start of the year Orion is high in the north (see Figure 13.10). From the southern hemisphere, Rigel is in the upper part of the constellation and Betelgeux at the lower right, with the Nebula above the Belt. Sirius is high, and therefore twinkles and flashes much less than it does as seen from Britain. Capella is low over the northern horizon; Crux is rising in the south-east, together with the Pointers.

Canopus (−0.7) is high; it is the leader of Carina, the Keel of the old ship Argo. Carina adjoins Vela (the Sails) and Puppis (the Poop), and the three take up most of the area between Crux and Sirius. The Milky Way is brilliant here, and there are clusters and nebulae as well as bright stars such as Beta Carinae (1.9). There is a very fine open cluster around Theta (2.8), and here too is Eta Carinae, the most erratic variable in the sky, which once outshone even Canopus but is now just below naked-eye visibility. It is associated with nebulosity, and is a fine sight even in a small telescope.

Note also the False Cross, which lies partly in Carina and partly in Vela. It is easily mistaken for the Southern Cross, but is larger, less brilliant and more symmetrical; its members are Epsilon Carinae (1.9), Iota Carinae (2.2), Delta Velorum (2.0) and

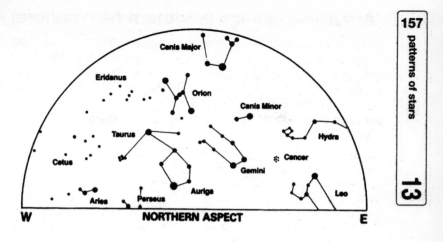

NORTHERN ASPECT

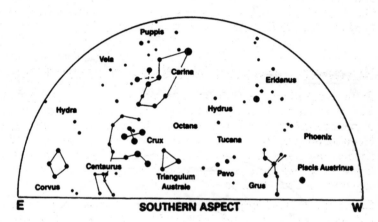

SOUTHERN ASPECT

Figure 13.10 Southern summer sky chart

Kappa Velorum (2.5). Like the real Cross, it has three white members and one – in this case Epsilon Carinae – which is orange.

Note the little constellation of Pictor (the Painter) near Canopus. It contains one star of special interest: Beta Pictoris (3.8), which may well be the centre of a planetary system.

Eridanus is a very long constellation, stretching from Achernar in the far south up to Orion, but it contains little of interest apart from the fine, easy double Acamar or Theta Eridani (2.9), the 'Last in the River'. Hydra occupies a large area of the eastern sky, and Leo is coming into view.

Autumn evenings (southern hemisphere)

Crux and Centaurus are now near the zenith, with Achernar very low down (see Figure 13.11). Canopus and the other stars of the Ship lie to the west; Orion has disappeared by late evening, but Sirius remains. Hydra sprawls across the zenith, with the quadrilateral of Corvus.

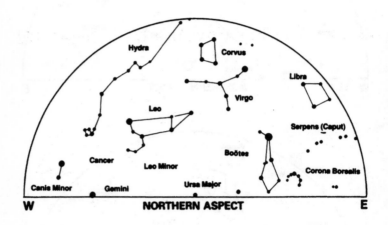

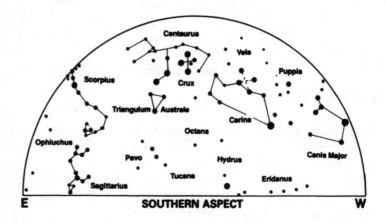

Figure 13.11 Southern autumn sky chart

Virgo, Boötes and Leo can be seen in the north. Scorpius is now prominent; it may be said to rival Orion in splendour, and the 'Sting' contains one star, Shaula or Lambda Scorpii (1.6) which is only just below the official first magnitude. Close by it is Lesath or Upsilon Scorpii (2.7). The two give the impression of being a very wide double, but they are not truly connected; Lesath lies in the background, and is four times as remote as Shaula.

Winter evenings (southern hemisphere)

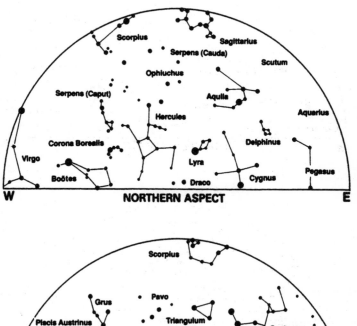

Figure 13.12 Southern winter sky chart

The scene is dominated by Scorpius, which is almost overhead; here too is Sagittarius, with its glorious star-clouds (see Figures 13.12 and 13.13). This is the best time of the year to see what northerners have come to call the Summer Triangle; Altair is reasonably high up, but Vega and (particularly) Deneb are always very low over the horizon. Arcturus is setting in the west, and much of the northern aspect is occupied by the large, dull groups of Hercules, Ophiuchus and Serpens.

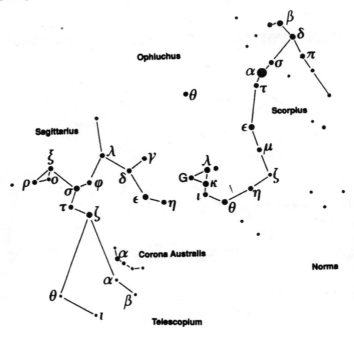

Figure 13.13 Scorpius-Sagittarius area

Crux is still high in the south-west, with Fomalhaut in the east; Canopus is so low that it will probably not be seen. High in the south we find the four Southern Birds: Grus (the Crane), Pavo (the Peacock), Tucana (the Toucan) and Phoenix (the Phoenix) (see Figure 13.14). This is a rather confusing area, and only Grus is at all distinctive. Of its two main stars, Alpha (1.7) is white; Beta (2.1) warm orange. The curved line of stars leading away from Alpha is easy to find; two of them, Delta (4.0) and Mu (4.8), look like wide doubles, though in each case we are dealing with line-of-sight effects.

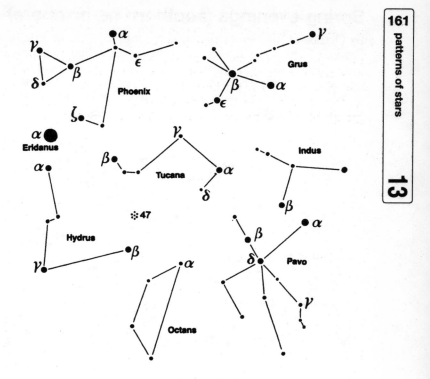

Figure 13.14 Southern Birds area

Tucana is redeemed by the presence of most of the Small Cloud of Magellan, together with the globular cluster 47 Tucanae. The cluster appears to be almost silhouetted against the Cloud, but is of course part of our Galaxy, whereas the Cloud is an independent system over 170 000 light-years away. Also high up is the Large Cloud, slightly closer to us and so bright that even strong moonlight will not drown it. It contains the great Tarantula Nebula, which is far larger than the Orion Nebula; in fact, if it were within a thousand light-years of us, it would cast shadows. Superficially the Magellanic Clouds look like detached portions of the Milky Way, but to astronomers they are among the most important objects in the sky.

Spring evenings (southern hemisphere)

By October, Orion is starting to come into view in the east (see Figure 13.15). Crux is very low, which means that Achernar is near the zenith; so too is Fomalhaut, which northern-hemisphere observers will find surprisingly bright. Canopus is rising in the east, with Scorpius and Sagittarius sinking in the west.

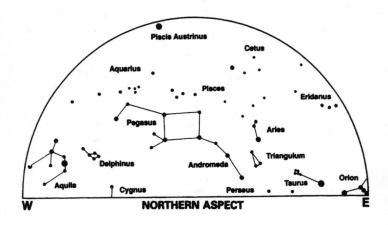

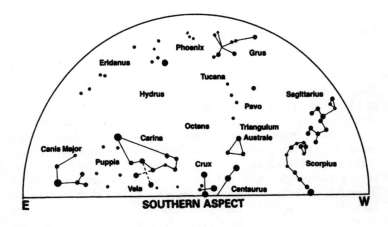

Figure 13.15 Southern spring sky chart

In the north Pegasus is on view, and leading away towards the horizon is the line of stars marking Andromeda. Of the 'Triangle', Vega and Deneb have disappeared, though Altair can still be seen low in the west.

I appreciate that this survey of the sky has been watered down to the point of dehydration, but it will suffice for the moment – and once the main patterns have been identified, it does not take too long to learn the rest. Remember, there are only a few thousand stars visible with the naked eye.

14

double and variable stars

In this chapter you will learn:
- about double stars
- about variable stars.

Now that we have taken a brief look at the sky as it appears throughout the year, it is time to say more about special types of stars. For example, there are stars which are double, or are members of groups; there are stars which change in light and heat output over short periods, and there are some stars which suffer violent explosions, so that they brighten up without warning and remain prominent for a few days, weeks or months before fading back into obscurity.

As we have noted, the most famous of all double stars is Mizar (Zeta Ursae Majoris), in the 'handle' of the Plough – or, if you like, the tail of the Great Bear. It is of the second magnitude, and close beside it is a much fainter star, Alcor (one of the few dim stars to have a generally-used proper name; officially it is 80 Ursae Majoris). Alcor is very easy to see under even average conditions, and it is genuinely associated with Mizar; even though the two are light-years apart, they share a common motion through space, and presumably have a common origin.

Telescopically, Mizar is itself seen to be double; the magnitudes of the components are 2.3 and 4.0, and they are separated by nearly 15 seconds of arc. Both are white, and both are more luminous than the Sun; the primary has 58 times the Sun's power, and the companion could match 12 Suns. They are moving together around their common centre of gravity, but they are at least 64 360 million km (40 000 million miles) apart, so that the revolution period is tens of thousands of years.

Mizar is a typical binary – that is to say, a physically associated system – but this does not apply to all doubles; we also have optical pairs, where one star is in the background, so to speak, and merely happens to lie more or less in the same line of sight as seen from Earth. Such is Theta Tauri, in the Hyades cluster near Aldebaran. The magnitudes of the components are respectively 3.4 and 3.8, so that they are not too unequal, and are separated by 338 seconds of arc, so that they can be seen separately with the naked eye; the brighter star is white, while the companion is orange. The stars are not bright enough for their colours to be well seen without optical aid, but binoculars show them excellently. In fact the orange star is almost twice as far away as its white companion, and there is no real connection between the two; if we were observing from a different vantage point in the Galaxy, the two Thetas could well be on opposite sides of the sky. It is rather surprising to find that optical doubles are not nearly so common as binaries – and it may even be that single stars such as our Sun are the exception rather than the rule. Stars tend to be gregarious.

Binaries such as Mizar have periods which are so long that the aspect does not change over several lifetimes, but in other cases the changes are obvious even over decades. With Gamma Virginis, in the 'bowl' of the Y of Virgo, the revolution period is only 171.4 years; the components are equal at magnitude 3.5. When I first looked at the pair, during the 1930s, any small telescope would show the components separately, but today they seem much closer together, and by 2010 they will be so close that Gamma Virginis will appear single except with giant telescopes. Of course, this does not mean that the components are genuinely approaching each other; it is simply that we are seeing them at a less favourable angle, and after 2020 the separation will increase once more. With Alpha Centauri, the brighter of the two Pointers to the Southern Cross, the revolution period is a mere 80 years, so that the separation alters quite quickly. This is also true of the position angle (P. A.), which is defined as the angle of the secondary from the primary star, from north (0 degrees) through east (90), south (180), west (270) and back to north. Amateurs can do useful work in measuring the separations and position angles of binary pairs, but accurate equipment is needed, and the procedure is not nearly so easy as it might sound.

Probably the loveliest double in the sky is Albireo or Beta Cygni, the faintest member of the cross of Cygnus. The primary is golden yellow, the companion vivid blue; the separation is 35 seconds of arc, and in any telescope, or even powerful binoculars, Albireo is a glorious sight. Some red stars are associated with companions which look green by contrast; such are Antares in the Scorpion and Rasalgethi in Hercules. Perfect twins are not uncommon; Gamma Virginis is the classic example, and another is Theta Serpentis, which lies not far from the boundary of Aquila. Each component is of magnitude 4.5, and the separation is over 22 seconds of arc. I always regard Theta Serpentis as one of the most attractive pairs in the sky.

In some cases the components of a binary pair are very unequal. Sirius, for example, has a companion which is very small but amazingly dense; it is a star of the type known as a white dwarf (I will say more about these in Chapter 15). The ratio in luminosity between the two is 10 000 to 1, though the companion is above the ninth magnitude and would be an easy telescope object if it were not so overpowered by its brilliant primary. The revolution period is 50 years. Small though it is, the companion is as massive as the Sun, and it is important to remember that the stars are not nearly so unequal in mass as they are in size and luminosity; the small stars are dense, the large stars rarefied. It is rather like balancing a lead pellet against a meringue.

If the separation between the two components of a binary is very small, no telescope will show the stars individually; but there is another method available, that of spectroscopic analysis. If the two stars are in motion around their common centre of gravity, one component must be approaching us while the other is receding. (Of course this effect must be superimposed upon the overall motion of the pair relative to the Earth, but this can be allowed for.) We can then make use of what is termed the Doppler effect. If a light-source is approaching us, the waves will be 'bunched' up and apparently shortened, so that all the spectral lines will be shifted over to the short-wave or blue end of the rainbow background; if the source is receding, the shift will be to the red. Therefore the approaching component of a close binary will have its lines blue-shifted while the receding component will be red-shifted, and all the lines in the spectrum will appear double; if both components are moving in a transverse sense, the lines will be single. Even if only one spectrum can be seen, the binary nature of the system will still betray itself, because the spectral lines of the visible component will oscillate around a mean position.

Both the components of Mizar are spectroscopic binaries (for that matter, so is Alcor), but an even better example is given by Castor, the senior though fainter member of the Twins. Visually, Castor is an easy binary; each component is itself a spectroscopic binary, and there is a third, much fainter member of the group which also turns out to be a spectroscopic binary – so that all in all there are six Castors, four bright and two dim. The view from a planet in such a system would indeed be spectacular.

Multiple stars can also be found. Epsilon Lyrae, close to Vega in the sky, is a naked-eye double, and a small telescope will show that each component is again made up of two, so that we have a double-double or quadruple system. Theta Orionis, in the Hunter's Sword, is nicknamed the Trapezium because of the arrangement of its four main components; and there are various other examples.

A binary system is not due to the splitting-up of a formerly single star, as was once thought; the components are born together, at the same time and in the same region of the same interstellar cloud. The components evolve differently because they are, usually, of different initial mass.

One particularly important binary is Algol, in Perseus. Normally it shines as a star of the second magnitude, but every 2½ days it begins to fade; it takes 5 hours to fall to below the third magnitude, remaining at minimum for a mere 20 minutes before taking another 5 hours to recover its lost lustre. It has long been known as the Demon Star (in the old constellation figure it marks the head of the Gorgon, Medusa, whose glance could turn any living creature into stone!), but apparently its odd behaviour was not noticed until the seventeenth century. It was explained in 1783 by John Goodricke, a young astronomer who was deaf and dumb – but who would certainly have achieved great things if he had not died before he was 23.

Algol is not truly variable at all. It is an eclipsing binary, made up of two components, one considerably brighter than the other. When the fainter star passes in front of the brighter, and hides well over 70 per cent of it, Algol seems to give a long, slow 'wink'. There is a slight fading when the primary eclipses the secondary, but the drop in brightness is too small to be noticed with the naked eye. The bright component is 100 times as luminous as the Sun, while the secondary is larger, cooler and less powerful. The distance between the two, centre to centre, is only about 11 million km (7 million miles), so that no telescope could show them separately even if they were equally bright.

Other Algol variables are known – Lambda Tauri, in the Bull, and the southern Zeta Phoenicis are other naked-eye examples – and there are eclipsing binaries of different type. Sheliak or Beta Lyrae, close to Vega, has components which are much less unequal than those of Algol, and are almost in contact, so that each must be stretched out into the shape of an egg. Variations are always going on, with alternate deep and shallow minima. The full period is almost 13 days, but we also meet with dwarf pairs – named after the prototype, W Ursae Majoris – where the periods are usually less than 24 hours.

At the other end of the scale we have Epsilon Aurigae, one of a triangle of faint stars close to Capella known collectively as the 'Haedi' or Kids. (Capella has been called the She-Goat.) The normal magnitude of Epsilon Aurigae is 2.9, and it has long been known to be a very luminous star, over 200 000 times as powerful as the Sun and at least 4500 light-years away. Every 27 years it fades by a magnitude, so that clearly it is being covered up, but the secondary has never been seen either visually or spectroscopically, and if it did not produce eclipses we would know nothing about it. It is probably a smallish, hot star surrounded by a vast cloud of gas and dust. The latest eclipse began in July 1982 and did not end until June 1984, so that nothing more can be expected yet awhile. Zeta Aurigae, also in the triangle of the Kids, is an eclipsing binary with a period of 972 days, made up of a large red star and a smaller blue star. It has no connection with Epsilon, and it is sheer coincidence that the two lie side by side in our sky; Zeta is little more than 500 light-years away from us.

Of the genuinely variable stars, the most useful to modern astronomers are the Cepheids, which are named after the first-discovered member of the class, Delta Cephei, in the far north of the sky – this discovery was, incidentally, made by the remarkable Goodricke. The period is 5.4 days, and the magnitude range is from 3.5 to 4.4. The period and the amplitude are absolutely regular, so that we always know what brightness Delta Cephei will be at any particular moment, and a light-curve can be drawn up by plotting magnitude against time. Cepheids are common enough, and several more are visible with the naked eye – notably Eta Aquilae in the Eagle, Zeta Geminorum in the Twins, and Beta Doradûs in the southern Swordfish.

In 1912 an American astronomer, Henrietta Swan Leavitt, was making careful studies of the two Clouds of Magellan, which lie in the far south of the sky and which look superficially like broken-off parts of the Milky Way. During her work, Leavitt discovered many Cepheid variables, and before long she noticed something very interesting. There was a link between period of variation and brightness; the longer the period, the brighter the star. Since the Clouds are so far away, it was good enough to say that all the stars in them are at the same distance from us, just as for most purposes we can assume that Charing Cross and Victoria Stations in London are the same distance from New York. It followed that the longer-period stars were actually the more luminous, and eventually it became possible to work out the power of a Cepheid merely by observing it. In fact, Cepheids are 'standard candles', and since they are highly luminous they can be seen over vast distances. They are pulsating stars, swelling and shrinking regularly. The period of pulsation is the time needed for a vibration to travel from the surface of the star through to the centre and back again, which explains why large, powerful stars have longer periods than smaller ones.

Of different type are the Mira variables, named after the first member of the class to be identified, Mira in Cetus (the Whale). It was seen in 1596 by a Dutch amateur, David Fabricius, when it was of the third magnitude. It then disappeared, and was not recorded again until 1603, when Johann Bayer noted it while he was compiling his famous star catalogue. It was again of the third magnitude, and Bayer gave it the Greek letter Omicron. Once more it vanished. Finally, in 1638, another Dutchman – Phocylides Holwarda – realized that it appears and fades away with fair regularity; the period is on average 331 days, and Mira is visible with the naked eye for only a few weeks in every year. At minimum it falls below the tenth magnitude, beyond the range of most binoculars and small telescopes.

Mira variables are very common indeed, though not many attain naked-eye visibility. All are red giants, with periods ranging from a few weeks up to several years; neither the periods nor the amplitudes are constant, and there is no convenient Cepheid-type period-luminosity law. Mira is much the brightest of them, and is usually visible with the naked eye for several weeks in every year. At some maxima it has been known to rise to magnitude 1.6, while at others it becomes no brighter than magnitude 4. Another well-known Mira star is Chi Cygni, in the cross of Cygnus, which has a much longer period (on average 407 days) and a greater range. At its best it has been known to reach magnitude 3.3, but at minimum it sinks to below 14, and since it lies in a very crowded region it is then hard to identify, despite its strong red colour.

Semi-regular variables are related to the Mira stars, and they too are red giants or supergiants, but they have very small amplitudes – seldom more than a couple of magnitudes, usually less – and have very rough periods interspersed with spells of chaotic behaviour. The most celebrated example is Betelgeux, in Orion. On rare occasions (as in late 1993 and early 1994) it can almost match Rigel, while at minimum it is little brighter than Aldebaran. The mean period is said to be 2110 days, but it is very rough indeed. Other conspicuous semi-regular variables are Beta Pegasi, in the Square; Eta Geminorum in the Twins, and Rasalgethi in Hercules.

Some variables are so unpredictable that we never know what they are going to do next. Such is R Coronae, in the 'bowl' of the Northern Crown, which is usually on the fringe of naked-eye visibility, but sometimes fades down to such an extent that a large telescope is needed to show it at all. Apparently it accumulates clouds of soot in its atmosphere, and these clouds cut off the light from below until they disperse. P Cygni, in the Swan, is a highly luminous and unstable star which may throw off shells of material; it has reached the third magnitude, but is generally just above the fifth. A less extreme shell star is Gamma Cassiopeiae, the central star of the W, which is subject to occasional outbursts that take it from below the second magnitude almost to the first. And in the far south of the sky we find the unique Eta Carinae, in the Keel of the Ship. At one time during the last century it outshone every star in the sky apart from Sirius, but for just over 100 years now it has been below naked-eye visibility. At its peak it shone as brilliantly as 6 million Suns, in which case it was the most luminous star in the Galaxy, and the total power is much the same even now, though most of it is in the infra-red part of the spectrum. It is associated with nebulosity, and when seen through a telescope looks quite unlike an ordinary star; I have described it as 'an orange blob'.

Novae, still often called 'new stars', are not really new at all, but are dim stars which suffer tremendous outbursts and flare up to many times their normal brilliancy before dying down again. A nova is a binary system, made up of one normal star together with a very old, very dense companion known as a white dwarf (see Figure 14.1). The white dwarf pulls material away from its

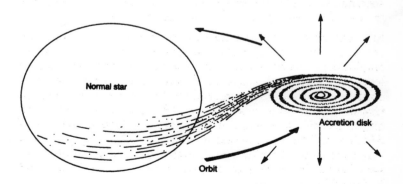

Figure 14.1 Theory of a nova

primary, and this material builds up until the situation becomes unstable; there is a sudden explosion, and the outburst may take weeks or even years to fade away. Some novae have been really spectacular; thus Nova Aurigae, seen in 1918, reached zero magnitude. Sometimes the outburst dies away quickly, as with the bright star seen in Cygnus in 1975, which flared up to magnitude 1.8 in a few hours but remained a naked-eye object for less than a week; on the other hand Nova Herculis 1934 was prominent for months, though it has now become very faint.

A few stars have been known to suffer more than one nova-like outburst, notably T Coronae in the Northern Crown, which is normally of the tenth magnitude, but which burst forth in 1866 and again in 1946 to reach magnitude 2. And if the build-up of material around the white dwarf becomes really extreme, the whole star may destroy itself in a truly catastrophic outburst. This is termed a type I supernova – but supernovae are rare, and the last in our Galaxy to become visible with the naked eye was seen as long ago as 1604.

Variable star observation has become one of the most important branches of modern amateur astronomy because there are so many variables that professional workers cannot possibly monitor them all. The procedure is to compare the variable with nearby stars of known brightness; with practice, eye estimates can be made accurate to a tenth of a magnitude, though specialized instruments known as photometers can naturally refine this further. Searches for novae are also carried out, and here the amateurs have a fine record of discovery. True, work of this kind is time-consuming and laborious, but there is always the chance of finding something new.

15

the life and times of a star

In this chapter you will learn:
- about stellar evolution for low-mass and high-mass stars.

A careful look into the night sky will show that the stars differ in colour as well as in brightness. True, most of them look white, but a few have a bluish cast, and some are clearly orange or orange-red.

The colours of the stars give a clue as to their surface temperatures; we all know that white heat is hotter than yellow, while yellow is hotter than red. The surface temperature of our yellow Sun is rather below 6000° C; the hottest known stars attain 80 000° C, while the coolest of the normal stars shine modestly at temperatures of 2500° C or even lower.

The astronomer's chief ally in stellar research is the spectroscope. As we have seen, the Sun – a typical star – yields a spectrum with a rainbow background (due to the relatively high-pressure gas of the photosphere) crossed by dark lines (due to the lower-pressure gases in the chromosphere). Normal stars show spectra of the same basic type, but there are marked differences in detail. For example, very hot stars have spectra dominated by lines of helium and hydrogen, while cool red stars show bands due to molecules of the sort which would be broken up at higher temperatures.

What has been done is to divide the stars into definite spectral types. The modern system was worked out at Harvard Observatory, and was intended to be alphabetical; the hottest stars would be of Type A, followed by B, C, D and so on. Inevitably there were complications, and some classes were found to be unnecessary, while others were out of order. The final sequence was alphabetically chaotic – W, O, B, A, F, G, K, M, R, N, S – but at least it led to the famous mnemonic 'Wow! O Be A Fine Girl Kiss Me Right Now Sweetie' (or, if you prefer, Smack). Two more classes, L and T, have now been added to deal with the very coolest stars; 'Like This!'.

Each type is again divided into sub-classes, so that, for example, A5 is half-way between A0 and F0. The main details are given in Table 15.1. Most of the stars are contained in the run from B to M; the first and last types are relatively rare, and R and N are now often combined as Type C.

Shortly before the First World War two astronomers, Ejnar Hertzsprung in Denmark and Henry Norris Russell in America, independently produced diagrams in which the stars were plotted according to their luminosities and their spectral types. It is clear that there is nothing random about the distribution. In a Hertzsprung–Russell or HR diagram, most of the stars lie on the band known as the Main Sequence, running from the top

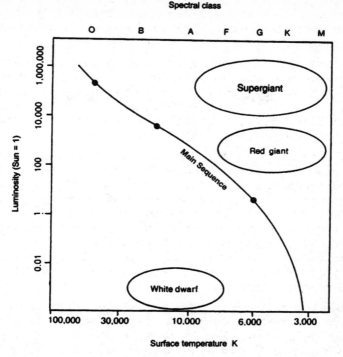

Figure 15.1 Hertzsprung-Russell diagram

left to the bottom right (see Figure 15.1). It is also clear that the orange and red stars are divided into two well-marked groups, one very luminous and the other very feeble; this giant and dwarf division is less marked for the yellow stars and does not apply to those which are white (the white dwarfs to the lower left come into a completely different category). Initially it seemed that we might be dealing with an evolutionary sequence. A star would begin by condensing out of the interstellar material, and would at first be large, cool and red, so that it would enter the HR diagram from the upper right. As it contracted, under the influence of gravity, it would become hotter and move over to the top left; it would then slide down the Main Sequence, ending up as a dim red dwarf before losing all its energy. Stars of types B and A were said to be of 'early' type; those of type M, 'late'.

It all seemed delightfully straightforward, but it was then found that the whole concept was wrong. The problem is that we

Table 15.1 Stellar spectra

Type	Colour		Surface temperature °C	Examples	Notes
W	White		Up to 80 000	Gamma Velorum (WC7)	Wolf-Rayet stars; bright lines Expanding shells
O	White		40 000–35 000	Zeta Puppis (O6)	Bright and dark lines
B	Bluish-white		Over 25 000–12 000	Rigel (B8)	Helium dominant
A	White		10 000–8000	Sirius, Vega	Hydrogen dominant
F	Yellowish-white		7500–6000	Procyon, Polaris	Calcium dominant
G	Yellow	giants 5500–4200		Capella	Metallic lines appear
		dwarfs 6000–5000		Sun	
K	Orange	giants 4000–3000		Arcturus	Strong metallic lines
		dwarfs 5000–4000		Tau Ceti	
M	Orange-red	giants 3400		Betelgeux	Bands due to molecules
		dwarfs 3000		Proxima	
R	Reddish		2600	No bright example	All very remote
N	Red		2500	R Leporis	Strong carbon lines
S	Red		2600	Chi Cygni	Almost all are variable

cannot (usually) see a star changing its evolutionary state, so that the only course is to pick out which stars are young and which are old. It transpired that far from being stellar infants, the red giants and supergiants such as Betelgeux are cosmical OAPs.

The change in outlook came with the realization that the stars shine because of nuclear transformations going on inside them, with hydrogen as the main 'fuel'. It then became possible to work out a proper sequence of events, and it emerged that the most important factor is the star's initial mass; massive stars develop much more rapidly than those of lower mass. Moreover, the range in mass is much less than in size or luminosity; there are very few stars more than 20 times as massive as the Sun.

We begin, as before, with the birth of a star from rarefied interstellar material – and we can see the first stages as 'Bok globules' (named in honour of the Dutch astronomer Bart Bok), which are black patches inside nebulae. Orion's Sword contains the great nebula M.42, which is a typical stellar nursery; inside it are many very youthful stars which have not yet become stable, and are varying irregularly. If the mass of the embryo star is less than about one-tenth of that of the Sun, the core will never become hot enough to trigger off nuclear reactions, and the star will simply shine dimly as a red dwarf until it has lost all its energy. With even lower mass, the temperature never rises above a few hundred degrees, so that we are really dealing with a sort of 'missing link' between a star and a planet; bodies of this kind are known, rather misleadingly, as brown dwarfs. The borderline between a planet and a brown dwarf is about ten times the mass of Jupiter, the senior member of the Sun's family.

If the mass is more than one-tenth of that of the Sun, the body attains true stellar rank. It begins to shine; as it shrinks the surface temperature remains the same, and the original cocoon of dust around the star's body is blown away. Irregular flickering continues, and there is a strong stellar 'wind'; this is the so-called T Tauri stage. Condensing to the Main Sequence takes millions of years, but at last the core temperature reaches the critical value of 10 million° C, and nuclear reactions start. The point on the Main Sequence where the star joins again depends on the mass; the greater the mass, the closer the point of junction to the upper left of the HR diagram (see Figure 15.1).

The hydrogen-into-helium conversion is adequate to maintain the flow of radiation for a very long time, but it cannot last for ever, and eventually it starts to run low; the core is now composed chiefly of helium. When the hydrogen supply fails, the star shrinks and the core heats up, so that helium can start to react and build up carbon; the star leaves the Main Sequence, and moves towards the giant branch at the upper right of the HR diagram. After a somewhat complex series of reactions it becomes a red giant, and the results so far as the Earth are concerned will be dire; for a while the Sun will radiate at least a hundred times as powerfully as it does now, and the outer layers will swell out until they engulf the orbits of the inner planets – Mercury and Venus certainly, and very probably the Earth as well. This means that life here cannot hope to survive, and indeed the Sun will have become uncomfortably hot even before then; recent calculations indicate that conditions on our world will become intolerable in little over 1000 million years hence. (Please do not worry. After all, it might be worse – say a mere 500 million years!) A star in this stage of evolution has become unstable, and is variable in output.

Next, the outer layers are blown away completely, and we have what is inappropriately termed a planetary nebula. The outer layers dissipate in space, and all that is left of the star is its core, which is now only a few tens of thousands of kilometres across, and is super-dense. The star has turned into a white dwarf, and has moved over to the lower left of the HR diagram.

The reason for this great density – sometimes over a million times that of water – is that the atoms in the star are crushed and broken. Under normal circumstances an atom is mainly empty space, but in the 'degenerate' matter of a white dwarf there is virtually no waste space at all, and the various components are packed tightly together. The best-known white dwarf is the faint companion of Sirius (often nicknamed the Pup, since Sirius is the Dog-star). The companion has only 1/10 000 of the luminosity of its brilliant primary, but it is as massive as the Sun, and a cupful of its material would weigh hundreds of tonnes.

A white dwarf has been aptly described as a bankrupt star; it has no reserves of energy left, and shines feebly only because it is still shrinking. In the end all its light and heat will leave it, and it will become a cold, dead black dwarf. Obviously we will then be unable to see it, because it will radiate no energy at all, but in any case it is by no means certain that the universe is yet old enough for any black dwarfs to have been produced. The universe, in its present form, is only 13 700 million years old at most, and the decline from a white to a black dwarf may take longer than that.

Now consider a star which is more massive still – say over 1.4 times as massive as the Sun. The birth process is the same as before, but everything happens at an accelerated rate, so that the star spends much less time on the Main Sequence before starting to run out of hydrogen fuel; it has of course joined the Main Sequence at the upper left. After the helium has reacted to form carbon, it is the turn of the carbon to start manufacturing heavier elements, and at one stage the star's structure is not unlike that of an onion, with different types of reactions going on at different levels. Iron is produced, and the core reaches the unbelievable temperature of around 3000 million° C.

It is when the core has been converted into iron that the real crisis begins, because iron will not react in the same manner as the lighter elements. Abruptly, all energy production stops. In a matter of seconds the core collapses; the outer layers fall on to it, and there is a tremendous rebound, so that a shock-wave spreads through the star's body and most of the material is hurled away into space in what is termed a supernova outburst. The luminosity may reach at least 5000 million times that of the Sun, and as the explosion dies down we are left with a cloud of material expanding into space; this material is enriched with the heavy elements which have been created during the build-up to the outburst, and it is out of this enriched material that new stars are formed.

But what of the core of the dying star itself? Even the component parts of the atoms are forced together; protons and electrons combine, and the positive charges of the protons cancel out the negative charges of the electrons, so that the result is a star made up of neutrons.

A neutron star is indeed bizarre. It is only a few kilometres across, but the density may be a thousand million times that of water, so that a pin's head of neutron star material would weigh as much as an ocean liner. According to theory, the outer surface of a neutron star is crystalline and iron-rich; below comes neutron-rich liquid material, which in turn overlays a core made up of particles about which we know almost nothing. There is a powerful magnetic field, and the neutron star is spinning around rapidly, perhaps many times per second. This tiny object is all that is left of the formerly very massive star.

In 1967 Jocelyn Bell Burnell, a radio astronomer at Cambridge University, England, made an unexpected discovery. She found a radio source which seemed to be 'ticking' so regularly and so rapidly that at first it was thought that the transmissions might be artificial. The 'LGM' or Little Green Men theory was soon discounted, but the pulsating source remained a mystery for some time. Finally, the cause was found. A 'pulsar' is a rapidly spinning neutron star; the magnetic axis is inclined to the axis of rotation, and beams of radiation are pouring out from the magnetic poles, so that every time a beam sweeps across the Earth we receive a pulse (see Figure 15.2). The situation is rather like that of a watcher on the coast being regularly illuminated by the beam of a rotating lighthouse far out to sea.

Many pulsars have now been found, but most of them are known only from their radio emissions, and only a few have been identified with optical objects. Of these the most famous lies in the middle of the Crab Nebula, an expanding gas-patch in Taurus which was discovered in 1731 by the English astronomer John Bevis; it lies near the third-magnitude star Zeta Tauri, and good binoculars will show it, though a telescope is needed to bring out its form – the nickname was bestowed on it in 1845 by the Earl of Rosse, who looked at it through his great 1.8-m (72-inch) reflector and thought that it gave a vague impression of a crustacean!

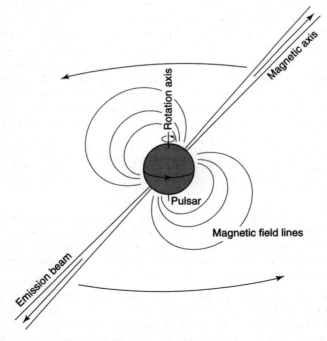

Figure 15.2 Pulsar diagram

We know a great deal about the history of the Crab because it is certainly the remnant of a supernova which blazed out in 1054, and was carefully followed by star-gazers in China and Korea. It became as bright as Venus, so that it could be seen in broad daylight, and it remained a naked-eye object for months before fading away. A pulsar was located in it during the 1960s, and was then tracked down to a very faint, flashing object which is visible only with sensitive equipment. There is no doubt about the identification, because both the 'flasher' and the pulsar have the same period; one-thirtieth of a second.

As a pulsar ages it slows down, and loses energy. This slowing-down can be measured, gradual though it is, and there are also 'glitches' or irregularities, due possibly to tremendous disturbances in the crust of the neutron star. Eventually, no doubt, the rotation will stop and the pulsar will cease to pulse, but again we are by no means sure that there has been sufficient time for this to happen. No pulsar is known with a period of more than five seconds.

To digress briefly: we also find pulsars which spin much more quickly than the Crab, at rates of hundreds of times per second. Apparently these are not young, but are old neutron stars which have been 'spun up'. A millisecond pulsar is thought to be the neutron-star component of a binary system, of which the other member is a huge, distended red giant. The neutron star pulls material away from the giant, and when the material hits the surface of the neutron star it increases the speed of rotation (you can demonstrate the effect by suspending a table-tennis ball on the end of a string and blowing on it through a straw). The giant suffers badly during this process, and in one case, a system in the constellation of Sagitta, we have a pulsar which is spinning 642 times per second and is actually evaporating its luckless companion, which will ultimately disappear completely. Not surprisingly, the attacking pulsar has been nicknamed the Black Widow!

Of course, not all neutron stars show up as pulsars, because we will not be able to detect the beams of radiation unless they sweep across us. Neither are we absolutely certain that all supernovae of this type produce pulsars, though all the evidence points that way.

Only four supernovae have been definitely recorded in our Galaxy during the past thousand years – those of 1006, 1054, 1572 (Tycho's Star) and 1604 (Kepler's Star), and of these only that of 1054 has left a pulsar. It is impossible to predict when we will see another; perhaps the best candidate is Eta Carinae, which is very massive and highly unstable and will definitely 'go supernova' eventually, though this may not happen for a great many centuries. At least a naked-eye supernova was seen in 1987, when it blazed out in the Large Cloud of Magellan and became very prominent even though it was 169 000 light-years away. We are still waiting for a pulsar to show itself, but so far nothing has happened.

Let us now come back to stellar evolution, and consider the case of a star which is too massive even to explode as a supernova. Here it seems that once the grand collapse starts, nothing can stop it; it is too cataclysmic. Gravitation takes over, and as the star becomes smaller and smaller, and denser and denser, the escape velocity rises, finally reaching the value of 299 300 km (186 000 miles) per second. This is the speed of light. Not even light can now break free from the star, which has surrounded itself with a 'forbidden zone' from which nothing – absolutely nothing – can escape. It has become a black hole (see Figure 15.3).

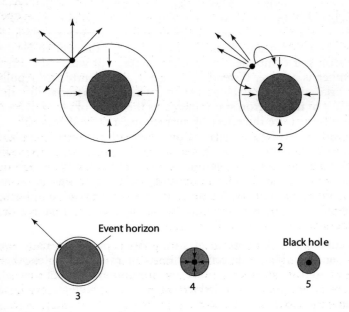

Figure 15.3 Formation of a black hole. 1. Light visible from surface as star begins to collapse. 2. As collapse accelerates some light rays are bent back. 3. As star approaches the critical radius, only vertical light rays escape. 4. When star collapses inside radius it vanishes from view. 5. Star falls into singularity.

The boundary of the black hole is termed the event horizon. If the Sun became a black hole, the diameter of the event horizon would be about 6.5 km (4 miles); for the Earth, less than 2.5 cm (1 inch). Yet there is no chance the Sun or the Earth will create black holes; even the Sun is not nearly massive enough, and is fated to end its luminous career much more sedately as a white dwarf.

Obviously we cannot see a black hole, but we can locate it indirectly. One candidate is a system known as Cygnus X-1, in the Swan, which is 6500 light-years away. It is made up of a very luminous B-type supergiant, with a diameter of at least 19.3 million km (12 million miles) and a mass 30 times that of the Sun, together with a companion which is 14 times as massive as the Sun; the companion has never been seen, but betrays its presence by its gravitational pull on the supergiant. The elusive companion is too massive to be a neutron star, and is presumably a black hole. It is drawing material away from the luminous star, and before this material is sucked over the event horizon, to vanish forever, it is so strongly heated that it gives off X-rays which we can pick up – hence the name Cygnus X-1. Confirmation has come, too, from the Hubble Space Telescope, which has been used to study the movement of material inside a galaxy known as M.87. The way in which this material is moving shows that it must be under the influence of an invisible body of immense mass, and this again seems to point to the existence of a black hole. Indeed, recent research with the Hubble Space Telescope indicates that with any large galaxy, a central black hole may be the rule rather than the exception.

A black hole is the weirdest object that we can imagine. What happens inside the event horizon we do not know; does the collapsed star crush itself out of existence altogether? Exotic theories – such as the idea of entering a black hole and emerging in a different part of the universe, or in a different universe altogether – are entertaining, but so speculative that further discussion of them is really rather pointless.

We have learned a great deal during the past few decades, and we are finding out more all the time. It seems that our theories about the life-stories of the stars are at least soundly based. There is one more very important point: do many stars have planetary systems of the same type as that of our Sun? And if the answer is 'yes', can these planets support intelligent life? I will have more to say about this in Chapter 18. At least there is plenty of time for life to develop. No star can shine for ever, but it takes thousands of millions of years to evolve from a mass of cool gas into a white dwarf, a neutron star or a black hole.

16

star-clusters and nebulae

In this chapter you will learn:

- about clusters of stars, both loose and globular
- about strange objects misleadingly called planetary nebulae, which are old stars that have thrown off their outer layers.

Look up into the evening sky at any time around Christmas, and you cannot fail to notice the 'haze' which makes up the star-cluster of the Pleiades, often known as the Seven Sisters. It has been known since prehistoric times, and almost every early civilization has legends about it. Closer inspection shows that it is starry, and at least seven separate stars can be seen with the naked eye under normal conditions; keen-sighted people can distinguish more, and the record is said to be 19. Binoculars bring many more into view, and the total membership of the cluster amounts to several hundreds. There is no doubt that the clustering is genuine; the chances of the main stars being arranged in this way by sheer chance are many millions to one against, and in any case many other open or loose clusters are known. Other prominent examples are the Hyades, round Aldebaran; Praesepe in Cancer, the Crab; the Sword-Handle in Perseus (not to be be confused with the Sword of Orion); the lovely 'Wild Duck' cluster in Scutum, and the equally lovely 'Jewel Box' around Kappa Crucis in the far south.

Open clusters cannot maintain their identity permanently. They are not tightly bound systems, and so they are constantly perturbed by field stars in the Galaxy which are not members of the group and are travelling in different directions. Eventually the cluster will be dispersed, though of course the process is a very gradual one, and the rate depends upon the position in the Galaxy. Thus Messier 67, in Cancer, is well away from the main galactic plane, so that it is in a fairly sparsely populated region, there are fewer field stars to disrupt it, and it is relatively old. Many of its stars are reddish, showing that they are well advanced in their life-stories and have left the Main Sequence; on the other hand the leaders of the Pleiades are hot, luminous and bluish-white, so that the age of the cluster cannot be more than a few millions of years.

Around a dozen clusters had been identified by the end of the seventeenth century, but the first really comprehensive catalogue was not drawn up before 1781. It was the work of a French astronomer, Charles Messier, who listed over 100 objects and numbered them; thus the Pleiades cluster is Messier 45 (M.45), Praesepe is M.44, and so on. Ironically, Messier was not in the least interested in clusters or nebulae. He was a comet-hunter, and found that he was being constantly misled by dim objects which lay well beyond the Solar System. Finally he lost patience, and decided to compile a list of them as 'objects to avoid'. We still use Messier's catalogue, though none of the comets he discovered during his career proved to be of any particular interest! Much later, the Danish astronomer J. L. E. Dreyer drew up a more elaborate catalogue, referred to as the NGC or New General Catalogue even though it is now over a century old. Both lists are in common use; thus Praesepe is known as NGC 2632 as well as M.44.

Some years ago I drew up a catalogue of bright clusters and nebulae which were not included in Messier's list and, rather to my surprise, this 'Caldwell Catalogue' has become widely used by amateur observers.

All these lists contain objects of various kinds, ranging from open clusters to globulars, nebulae, galaxies and even supernova remnants (the Crab Nebula is M.1 or NGC 1952). All the objects in the Messier and Caldwell catalogues are within the range of small telescopes, and identifying them is a pleasant pastime; though they are not easy to draw accurately, they are favourite targets for astronomical photographers. Of course they differ widely in size and prominence. Some of the clusters are rich, while others are poor and obscure; there are two Messier objects – numbers 40 and 91 – which are missing and may be comets which were not identified at the time, while M.102 is generally thought to have been identical with M.101, a spiral galaxy in Ursa Major.

Large open clusters such as the Pleiades and the Hyades are probably best viewed with binoculars, because they are too spread out to be contained in the same telescopic field. The Pleiades cluster contains nebulosity, showing that there is still 'star-forming' material present; this nebulosity is not hard to photograph, but is very elusive visually. There is no nebulosity in the Hyades cluster, which was not listed by Messier – presumably because there was not the slightest chance of confusing it with a comet. Another absentee from Messier's catalogue is the Sword-Handle in Perseus, made up of two clusters in the same field. M.11, in Scutum, is fan-shaped, and has been nicknamed the 'Wild Duck', while Cancer contains two prominent clusters: M.44 (Praesepe), an easy naked-eye object, and M.67, which can be seen with binoculars. Praesepe, the 'Beehive', is bounded by two naked-eye stars, Delta and Gamma Cancri, which are known as the Aselli or Asses, since another nickname for Praesepe is the 'Manger' (it is not easy to see why the ancient Chinese referred to it as 'the exhalation of piled-up corpses'!). As we have noted, M.67 is exceptionally ancient by open cluster standards because it lies well away from the plane of the Galaxy and has been left relatively undisturbed by passing non-cluster stars. In the Southern Cross we find the exquisite Jewel Box, with bluish-white stars and one prominent foreground red giant, while there are many other clusters within binocular range.

Open clusters have no particular shape, and seldom contain more than a few hundred stars. Globular clusters are quite different (see Figure 16.1). They are huge, symmetrical systems, sometimes with more than 1 million members, lying around the edges of the main Galaxy. Only about 100 are known in our own system, and only 3 are clearly visible with the naked eye: Omega Centauri and 47 Tucanae in the far south, and M.13 in Hercules. All globulars are very remote, which is why they look comparatively faint.

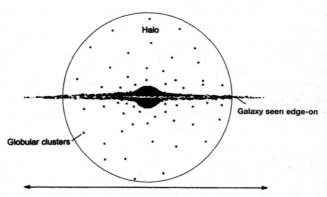

Figure 16.1 Distribution of globular clusters

Globular clusters are old, and have no nebulosity left. Their leading stars tend to be ancient red giants which have long since left the Main Sequence. However, there are also Cepheid variables, and it was by studying these useful stars that Harlow Shapley, in 1917, was able to draw up the first good picture of the shape of the Galaxy. He could observe the periods of the Cepheids, and find out how far away they were; this meant that he also knew the distances of the globular clusters in which they lay. He also realized that the globulars are not distributed evenly around the sky, but are concentrated in the southern hemisphere, particularly in the region of Sagittarius. Shapley deduced – quite correctly – that this is because the Solar System lies not at the centre of the Galaxy, but well out towards one edge, so that we are having what amounts to a lop-sided view.

We now know that the Sun is about 27 000 light-years from the galactic centre, which lies beyond the Sagittarius star-clouds.

Telescopically, it is easy enough to resolve the edges of globular clusters into stars, but near the centre the stars are so closely packed that they seem to merge into a general mass; the average distance between stars in the heart of a globular is only a few light-days instead of several light-years. The sky as seen from a planet moving around a star in such a region would be ablaze, with many stars brilliant enough to cast shadows; there would be no proper darkness at all, and of course many of the brightest stars would be red. Yet an astronomer living near the centre of a globular cluster would be able to find out very little about the outer universe!

Much the brightest of the globulars is Omega Centauri, which shows up to the naked eye as a hazy object of about the fourth magnitude; it lies in the northern part of the Centaur (declination – 47½ degrees) and Johann Bayer gave it a Greek letter. It is one of the closer members of the class, but even so it is still 17 000 light-years away; the real diameter cannot be less than 150 light-years. Unfortunately it is too far south to be seen from Britain, and this is also true of its only rival, 47 Tucanae, which is almost silhouetted against the Small Cloud of Magellan. The best northern globular is M.13, between Eta and Zeta Herculis, which is a difficult naked-eye object but is very easy with binoculars, and was discovered in 1714 by no less a person than Edmond Halley. Its outer parts are easy to resolve, but oddly enough it is relatively poor in variable stars.

The most detailed views of the central regions of globular clusters have been obtained with the Hubble Space Telescope. In particular, it has been possible to identify what are termed 'blue stragglers', hot blue stars which logically ought not to be there; globular clusters are so ancient that all their leading stars should have left the Main Sequence by now. Apparently a blue straggler is formed when two stars approach each other closely enough to form a binary system, after which the more massive member draws material away from its more evolved, less dense companion – with the result that the main star heats up, becoming a blue giant again. Over 20 blue stragglers are known inside 47 Tucanae, of which 11 were found with the Hubble Telescope in 1991. However, the results did not confirm a popular suggestion that lurking in the centre of a globular cluster there is likely to be a super-massive black hole.

Messier's catalogue includes several of the objects always known as planetary nebulae, though as we have noted the name is a bad one inasmuch as the objects concerned are not true nebulae and have absolutely nothing to do with planets. A planetary nebula is a very old star which has blown away its outer layers, so that it is surrounded by an extensive, tenuous shell. When we look at this shell we see it at its brightest near the edges, giving the aspect of a ring around a small, very hot central star. The best-known planetary is the Ring Nebula, M.57, between Beta and Gamma Lyrae, near Vega. It is hard to see with binoculars, but a small telescope shows it, and a larger instrument will reveal the central star; the overall appearance has been likened to a tiny, luminous cycle-tyre. Not all planetaries are as symmetrical as this; M.27, in Vulpecula (the Fox), earns its nickname of the Dumbbell, while two stars in the much fainter M.97, in Ursa Major, have led to its being generally known as the Owl.

One planetary nebula, Hen 1357, has actually been observed in the process of formation. (It is so called because it was the 1357th object in a list of unusual stars compiled by Karl Henize.) It is about 18 000 light-years away, and lies in the southern constellation of Ara, the Altar. When first noted by Henize it looked like an ordinary hot star, but in 1993 a picture taken by the Hubble Space Telescope revealed that it had blown away its outer shell and had taken on the aspect of a planetary nebula. In fact, the process may have started several thousands of years ago, but only recently has there been enough radiation from the star to make the ejected gas glow.

The Crab Nebula is the only supernova remnant in Messier's list (appropriately, it is lettered No.1), but there are a number of gaseous or galactic nebulae. The best-known is M.42, the Sword of Orion (see Plate 11), which is 1500 light-years away and about 30 light-years in diameter. It is a stellar nursery, and contains many very young stars which are still contracting towards the Main Sequence, so that they are flickering irregularly. We cannot see through to the heart of M.42, but infra-red radiation is not blocked by the nebular material, and we know that deep inside the 'Sword' there are some massive young stars which we will never be able to observe directly; they will not live for long enough to 'burn a hole' and show themselves.

Nebulae shine because they are being lit up by hot stars in or very near them. Some nebulae shine purely by reflection, but in other cases the very hot stars ionize the hydrogen and make it emit light on its own account, which is why these emission nebulae are also termed H.II regions. M.42 is only the brightest part of a vast molecular cloud covering most of Orion. The four stars of the Trapezium, Theta Orionis, lie near the edge of the Earth-turned side of the nebula, and it is these stars which make the 'Sword' shine. Among other conspicuous nebulae are the Lagoon (M.8) and the Trifid (M.20) in Sagittarius, and the nebulosity surrounding the erratic southern variable Eta Carinae.

But suppose that there are no convenient stars to make the material shine – what then? The answer is that the nebulosity will remain dark, and will be detectable only because it blots out the light of stars beyond. The most famous of all dark nebulae is the Coal Sack, in the Southern Cross, which covers almost 30 square degrees and looks like a starless area (or nearly so; there are a few faint foreground stars showing up against the blackness). Another dark nebula, near Alnitak in Orion's Belt, is known as the Horse's Head because its shape recalls that of a knight in chess; it is very hard to see visually; even with a large telescope, but it is not difficult to photograph, and it stands out clearly against the bright nebulosity surrounding it. In Cygnus we can locate dark rifts in the Milky Way, and these too are due to obscuring material.

There is no difference between a bright nebula and a dark one – except for the presence or absence of illuminating stars. For all we know, there may be a 'Trapezium' on the far side of the Coal Sack which will light up the regions turned away from us; if we were observing from another point in the Galaxy, it might be the Coal Sack which would look bright and Orion's Sword which would remain dark. At least there is plenty of variety in the Galaxy!

17

the depths of the universe

In this chapter you will learn:
- about the galaxies, vast star-systems many millions of light years away
- how the groups of galaxies are moving away from each other so that the whole universe is expanding.

In this light-polluted era there must be many people who have never seen the Milky Way, but against a clear, dark sky it is magnificent, and it must have been known ever since the dawn of human history; every country has a store of legends about it. It was Sir William Herschel who first set out to determine the shape of the Galaxy, and he who realized that the Milky Way is nothing more than a line-of-sight effect. He came to the conclusion that the Galaxy must be shaped like 'a cloven grindstone', which is not so very far from the truth, but its size was not measured until Harlow Shapley carried out work from Mount Wilson in 1917.

As we have noted, the Sun lies about 25 000 light-years from the galactic centre, and we also know that the Galaxy is rotating; the Sun takes 225 million years to complete one circuit – a period often called the 'cosmic year'. One cosmic year ago, the most advanced life-forms on Earth were amphibians; even the dinosaurs lay in the future. It is interesting to speculate as to what conditions on Earth may be like one cosmic year hence!

What we cannot do is see right through to the centre of the Galaxy. We know where it is; it lies beyond the lovely star-clouds in Sagittarius, but it is hidden by the 'dust' which lies in the way. Fortunately, there are other methods of investigation. Infra-red radiation is not blocked by the dust, and neither are radio waves, so that we can learn at least something about conditions in the heart of the Galaxy. There is a strong radio source, known as Sagittarius A-* (pronounced Sagittarius A-star), which is a massive black hole.

We also know that the Galaxy is spiral in form. The first evidence here came from radio astronomy, not long after the end of the Second World War. There are clouds of cold hydrogen spread through the Galaxy, and these give off radio emissions at a wavelength of 21.1 cm (8.31 inches). Plotting the distribution of these clouds confirmed that there are spiral arms, and that the Sun lies near the edge of one of them. There is nothing at all surprising in this, and neither is there anything exceptional about our own Galaxy.

How far can you see with the naked eye? Asked this question, most people will reply 'Oh! – a few miles.' In fact the answer is 'around 19.3 million million million km' (12 million million million miles), because this is the distance of M.31, the Andromeda Spiral (see Plate 12), which is about 2.2 million light-years away and is the most remote object which can be clearly seen without optical aid. But for a long time its nature was not realized.

The nebulae in Charles Messier's catalogue were of two distinct types. Some, such as M.42 in Orion's Sword, looked like masses of gas; others, such as M.31, gave every impression of being 'starry'. Nobody could be sure whether the starry nebulae were members of the Milky Way, or whether they were independent systems – 'island universes' – at much greater distances. In 1845 the third Earl of Rosse, using his home-made 1.8-m (72-inch) reflecting telescope at Birr Castle in Ireland, saw that many of the starry nebulae were spiral in form, like Catherine-wheels, though others were circular, elliptical or even irregular in shape. Certainly they were quite different from the gas-patches, but everything depended upon how far away they were.

The problem was solved in 1923 by Edwin Hubble, again using the Mount Wilson 2.5-m (100-inch) telescope. He knew about the Cepheid period-luminosity law established years earlier by Henrietta Swan Leavitt, and he set out to locate Cepheids in some of the spirals, including M.31. He found them, and measured their periods; this gave him their distances, and it was at once clear that they could not possibly be contained in our Galaxy. Hubble gave the distance of M.31 as 900 000 light-years, subsequently reduced to 750 000 light-years. This was one of the greatest discoveries in the history of astronomy, and it altered all our ideas about the universe. Just as the Earth had been found to be an unimportant planet and the Sun an unimportant star, so it now emerged that the Milky Way itself is an unimportant galaxy.

More was to follow. In 1952 Walter Baade used the then-new Palomar 5-m (200-inch) reflector to show that there had been an error in the Cepheid scale. There are two kinds of short-period variables, one much more luminous than the other, and Hubble, through no fault of his own, had picked the wrong one. The variables which he had studied were of the more powerful type, and so the distance which he had given for M.31 was an underestimate; it had to be increased by a factor of more than two, and there were similar errors in the distance of the other galaxies. In one short paper, delivered at a meeting of the Royal Astronomical Society in London, Baade calmly doubled the size of the universe.

But nothing could detract from the value of Hubble's work; it was he who made the great breakthrough. Together with his colleague Milton Humason (who began his Mount Wilson career as a mule driver, graduating to the rank of janitor and then to the status of one of the world's greatest astronomers!),

Hubble drew up a classification of the external systems, based on their forms. The old term of 'spiral nebulae' was dropped, to be replaced by 'galaxies', while 'nebula' was taken to indicate a gas-and-dust cloud.

Hubble divided the galaxies into five main classes. First there were the spirals, some of which are truly beautiful; it is easy to see why M.51, in the little constellation of Canes Venatici (the Hunting Dogs) has been nicknamed the Whirlpool. M.31 is also spiral, but it lies at a narrow angle to us, so that the full effect is lost. There are various subdivisions, lettered a, b and c. Sa spirals have conspicuous, often tightly wound arms issuing from a well-defined nucleus; Sb systems have looser arms and less condensed nuclei, while in the Sc spirals the nucleus is inconspicuous and the arms are very loose indeed. Our Galaxy is of type Sb.

With barred spirals, the arms issue from the ends of a kind of 'bar' through the nucleus; they are subdivided into types SBa, SBb and SBc. Elliptical galaxies show no signs of spirality; they range from E7 (highly flattened) through to E0 (virtually spherical, and looking superficially very like globular clusters even though they are far more populous and massive) (see Figure 17.1). Finally there are the irregular galaxies, with no definite form. It is now thought that about 30 per cent of galaxies are spiral, 60 per cent elliptical and 10 per cent irregular, though these figures are approximate.

It was natural to think that spirals should evolve into ellipticals, or vice versa, but this does not seem to be true. More probably the final form depends upon the initial mass of the galaxy concerned. Neither does it seem that spiral arms are permanent. 'Density waves' sweep around the system and compress the material, triggering off star formation; this means that in a

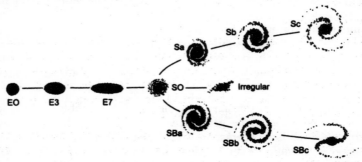

Figure 17.1 Hubble's tuning-fork diagram

spiral galaxy the arms will always exist – but they will not always be the same arms, or in the same positions.

It cannot honestly be said that galaxies are spectacular objects when observed visually, even in large telescopes; the best results are obtained photographically, or today with electronic equipment. However, we can at least see that the galaxies contain all the various types of objects with which we are familiar in the Milky Way. There are open and globular clusters, planetaries, gaseous nebulae, variable stars and novae; supernovae flare up from time to time, and their immense power means that they can be seen across great distances. In recent years some amateur astronomers have concentrated upon hunting for supernovae in outer galaxies, and have a fine record of success.

Our Galaxy is a member of what is termed the Local Group, which includes a few large systems and more than two dozen of smaller size. The closest of the important systems are the two Clouds of Magellan, which are less than 200 000 light-years away, and so can be studied in great detail. At a greater distance come M.31, the Andromeda Spiral (see Plate 12), and the smaller spiral M.33 in Triangulum.

The junior members of the Local Group are real dwarfs, some of which are hardly more populous than globular clusters and have no definite shape. The closest of them – and, indeed, the closest of all the external systems – is a dwarf in Sagittarius, discovered in 1994 by Rodriga Ibata and his colleagues working at Cambridge, England. It is about 80 000 light-years away from us, but lies on the far side of the Galaxy, and is no more than 50 000 light-years from the galactic centre. It – like the Clouds of Magellan – is a satellite of our Galaxy, but unlike the Clouds it is being gravitationally disrupted, and over the next few hundred million years it will be absorbed into our Galaxy and will lose its separate identity.

The Local Group is stable, and indeed M.31 is actually approaching us at the present time. Collisions can, and do, occur, and the Hubble Space Telescope has found indications of a double nucleus in M.31 which may indicate the remnant of a smaller system that was absorbed long ago – though this is only one interpretation, and we cannot be sure.

Beyond the Local Group we come to other groups or clusters of galaxies, many of them far larger than ours; the Virgo cluster, at around 50 million light-years, contains many hundreds of members, notably the giant elliptical M.87, which is also a strong source of radio waves (radio astronomers call it Virgo A). The Virgo cluster is probably the hub of a vast collection of systems known as the Local Supercluster. But all the groups are racing away from us, and from each other. This means that the entire universe is expanding.

Well before Hubble began his work, Vesto Slipher at the Lowell Observatory in Arizona had made an important discovery. He found that apart from the systems contained in what we now know to be the Local Group, all the galaxies showed red shifts in their spectra and, assuming that these were Doppler effects, it followed that all the galaxies were receding. When Hubble measured their distances, it became clear that we are in no privileged position, and the expansion really is universal.

Cepheids are invaluable in measuring the distances of the closer galaxies, but with more remote systems the individual variables fade into the general background, and we must think again. We can use supergiant stars, assuming that the most luminous stars in other galaxies are of around the same power as the most luminous stars in the Milky Way; we can also use supernovae. But eventually even the supernovae are lost, and we come back to the spectral red shift. Find the speed of recession, and from this make an estimate of the distance. Though uncertainties are bound to creep in, it seems that the most remote systems known are at least 13 000 million light-years away, and are receding at well over 90 per cent of the velocity of light.

There have been well-known astronomers who have questioned this whole interpretation. In particular Halton Arp, one of the leaders in studies of galaxies, drew attention to galaxy-galaxy, quasar-quasar pairs which seemed to be connected, in some cases by bridges of luminous material, and which however add completely different red shifts. He concluded that the red shifts were not due purely to the Doppler effect, and that therefore all our measurements of distances beyond the Milky Way were wrong; quasars would be minor features ejected from galaxies. A strong supporter of this interpretation was Sir Fred Hoyle. It is true that Arp's observational results were difficult to reject as being due to sheer chance alignments (they still are!), but very few astronomers today believe that the red shifts are anything but cosmological.

By 1963 radio astronomy had become a vital part of research. Bernard Lovell's 76-m (250-ft) 'dish' at Jodrell Bank was in full operation, and catalogues of radio sources had been drawn up, notably by the Cambridge (England) teams. The main problem was in identifying the sources with optical objects; in those days radio telescopes were not nearly so accurate as they are now. For once nature was helpful. A strong source, 3C-273 (the 273rd object in the third Cambridge catalogue) was occulted by the Moon, and this gave its position very precisely, so that it could be tracked down to what looked like a faint blue star. When the optical spectrum was examined, initially by Maarten Schmidt at Palomar, California, astronomers received a rude shock. 3C-273 was not a star at all, but something much more dramatic. The spectrum was totally unstellar, and showed hydrogen lines with a tremendous red shift. This meant that the object was very remote and immensely powerful; yet it looked small. How could so much energy come from so limited an area? If the red shift gave a real key to the distance, this first 'quasar', or quasi-stellar object, was much more powerful than any normal galaxy.

Other quasars were soon found, and by now thousands are known, though by no means all of them are radio emitters. They are the nuclei of very active galaxies, powered by super-massive central black holes, and it may be that many galaxies go through a temporary quasar stage during their evolution; there is almost certainly a gradation between quasars, the so-called BL Lacertae objects (which are rather less luminous) and other galaxies which are quite different in appearance from quasars, but are very powerful at radio wavelengths. It seems that the three classes of objects may be of the same type, seen from different angles.

Quasars are so luminous that they can be seen at distances where normal galaxies have faded into invisibility. In some cases they show multiple images – not because they are really made up of separate components, but because their light passes close by a foreground galaxy which acts as a lens, and produces several images of the same quasar. Such is the so-called 'clover-leaf ', where there are four images of the background quasar surrounding the intervening galaxy which is responsible for the lens effect.

All this means that we can now study the distribution of the galaxies in space, and initial results have been decidedly unexpected. There is a definite large-scale structure, with 'walls', super-clusters, and also vast voids which are only sparsely populated. But there is also a tremendous amount of material which we cannot see, and this brings us on to one of the most pressing problems of modern cosmology: that of the so-called missing mass.

A galaxy such as our own is rotating, with the individual stars moving around the centre of the system. According to Kepler's Laws, the closer-in stars should move the fastest, as with the planets in the Solar System; but this does not happen, and the orbital speeds do not fall away in Keplerian fashion. The only conclusion must be that the mass of the galaxy is not concentrated in the centre of the system, and that there is a great deal of invisible material. This also explains why the clusters of galaxies retain their separate identities; there is insufficient visible matter to make them stable, and so 'dark matter' must be involved.

It has even been calculated that the visible material in the universe – planets, stars, galaxies, everything – accounts for no more than 10 per cent of the total mass. All the rest is undetectable.

What can it be? Is it locked up in innumerable black holes? Are there swarms of stars so feeble that they do not show themselves? Or is the dark matter of a nature completely alien to us, so that our present-day equipment has no hope of detecting it? As yet we do not know, but the problem is of fundamental importance when we come to consider the past and future of the universe. In particular, we want to decide whether the present expansion will continue indefinitely, or whether there is sufficient mass to bring the expansion to a halt; and this leads on to the question of how the universe was created in the first place.

There are two alternatives. Either the universe was born at a definite moment in time, or else it has always existed, so that it has an infinite past. Neither concept is at all easy to grasp. We cannot visualize a period of time which had no beginning, and in any case this is no help in trying to understand how the material came into being; the one inescapable fact is that we exist, so that the atoms and molecules making up everything in the universe must have come from somewhere or other. If we adopt the other alternative, and assume that the universe began with a 'Big Bang' 13.7 thousand million years ago, we have to ask ourselves what happened at a still earlier period. All we can really say is that if space, time and matter all began at the same instant, there was no 'before'.

If we begin with a Big Bang we can work through a complete sequence of events, ending up with you and me. The universe was born; it was very small and incredibly hot, but we cannot say 'where' the Big Bang happened, because if space was created at the same moment then the Big Bang happened 'everywhere'. Expansion began – very rapidly at first, in what is termed the inflationary period – and then more slowly. The temperature fell; complex atoms formed from the original hydrogen, and in time galaxies were produced, followed by stars, planets, and – in the case of the Earth – life.

Further evidence comes from weak radio emissions reaching us from all directions all the time, indicating an overall temperature for the universe of about three degrees above zero (absolute zero being the coldest temperature there can possibly be: $-273°$ C). Presumably this is the last detectable manifestation of the Big Bang. If the original universe had begun to expand smoothly, it is not easy to see how galaxies could have been formed, but in 1993 a specially designed satellite, COBE (the Cosmic Background Explorer) was able to detect slight irregularities in the weak radiation, so that another major objection to the Big Bang theory was removed.

When we ask just how the Big Bang was caused, we are forced to admit our total ignorance, because all the ordinary laws of science break down. I have compared this with the plight of an intelligent being from outer space who arrives on Earth for a brief visit and spends half an hour in a busy street. He will see babies, boys, men and old men. If he is clever enough, he will realize that a baby turns into a boy and a boy into a man, so that he will be able to work out the evolutionary cycle of a human being – but unless someone has told him about the facts of life, he will not know how the baby appeared. In cosmology, our 'baby' is the Big Bang.

Whether the overall expansion will continue indefinitely is something else which we do not know, and again we come back to the question of the average density of matter in the universe. If the mass exceeds a certain critical value, the expansion will stop; the galaxies will start to rush together again, and in perhaps 80 000 million years there will be another Big Bang, so that the cycle can be repeated. If the total mass is too low, then there will be no return, and the expansion will go on until all the groups of galaxies have lost contact with each other. In the first scenario, the universe resembles a clock which is being regularly re-wound; in the second, it is like a clock which has been wound once and is now running down, so that eventually it will stop.

New data about the very early universe were obtained in 1999 from the project known as BOOMERANG (Balloon Observations of Millimetric Extragalactic Radiation and

Geophysics). The main telescope had a 1.2-m (47-inch) primary mirror, and the equipment was carried in a helium-filled balloon which flew around Antarctica in December 1998 to January 1999 – at an altitude of 37 km (23 miles) – Antarctica being chosen because of the stable high-altitude winds and the constant sunshine. The equipment could measure temperature variations down to 0.0001 of a degree – far better than COBE. The results supported the earlier results.

It is natural to assume that following the Big Bang, and the brief period of rapid inflation, the rate of expansion of the universe should slow down, because of the effects of gravity. However, it now seems that the rate of expansion is actually accelerating with great distance. Very remote supernovae, which act as 'standard candles' because they all reach the same peak luminosity, appear fainter than they should do, so presumably they are further away than expected – because of an increased expansion rate (in fact, to be really accurate, it is space itself which is expanding, carrying the galaxies with it). There could even be a mysterious force of repulsion, acting against gravity. This idea had been introduced by Albert Einstein, though he later abandoned it. All this seems to indicate that the universe is 'open' rather than closed, so that expansion will never stop, but the evidence is not conclusive, and there may be many more factors to be taken into account. The jury is still out.

We are equally insecure when we consider the full size of the universe. Is it 'finite but unbounded'? However, we do have one clue. If the rule of 'the further, the faster' holds good, we will come to a distance at which a system will be receding at the full velocity of light; we will then be unable to see it, and we will have come to the boundary of the observable universe, though not necessarily of the universe itself. It seems that the limiting distance is 13.7 thousand million light-years. More powerful equipment now being planned – for example, a new Space Telescope, the James Webb Telescope, larger than Hubble – may take us to the 'outer limit', but for the moment we can do no more than wait and see.

We have come a long way since the time, not so very long ago, when the Earth was regarded as the centre of the entire universe, with everything in the sky produced for our sole benefit. Yet we are still woefully weak on fundamentals, and we cannot pretend that we have made much progress in understanding the most important problem of all – that of the Creation. One day, perhaps, we may find out.

18
into the future – life beyond the Earth

In this chapter you will learn:
- about new investigations into the possibility of life on other worlds.

In 1991 a General Assembly of the International Astronomical Union, the controlling body of world astronomy, was held at Buenos Aires, in Argentina. One meeting was of unusual interest. It dealt with the possibility of ETI, or Extra-Terrestrial Intelligence. Does it exist, and what action should we take if we found any evidence of it? After all, the Sun is a very normal star, one of 100 000 million in our Galaxy alone, and it does not seem logical to believe that it is unique in being attended by a peopled planet. But we must beware of jumping to conclusions, and we first have to decide what form any ETI is likely to take.

We may not know a great deal about 'life', but we do know a great deal about the properties of living matter, and it seems that the essential ingredient is carbon, because only the carbon atom has the ability to link up with other atoms to form the large, complicated molecules necessary for life. (Silicon is the only possible competitor, but it is not very efficient, and the chances of silicon-based life anywhere seem to be low.) We are therefore entitled to assume that life, wherever it may be found, will be carbon-based. If this is wrong, then almost all of our modern science is wrong too, and few people will be prepared to accept this.

If there is a planet like the Earth, orbiting a star like the Sun, there might well be Earth-type life; and if it has developed in the same way as our own, there could be a distinct chance of contacting it. Interstellar travel is out of the question in our present state of technology, if only because it would take far too long, and when we consider devices such as interstellar arks, space-warps, time-warps, teleportation and thought-travel we enter the realm of science fiction. So if we are to establish contact with other civilizations, the only feasible way seems to be by radio, since radio waves are electromagnetic vibrations and travel at the same speed as light.

The first serious attempt was made in 1960 by astronomers at Green Bank, in West Virginia. Using the powerful radio telescope there, they 'listened out' at a selected wavelength, 21.1 cm (8.31 inches), hoping to pick up signals rhythmical enough to be interpreted as artificial. This wavelength was chosen because it is the frequency sent out by clouds of cold hydrogen in the Galaxy, and radio astronomers anywhere would be expected to pay particular attention to it. Not surprisingly, the results were negative, and the experiment was discontinued; later attempts were no more successful. Yet although the chances of picking up artificial transmissions are slight, they are not nil.

However, the first step was to show that planets of other stars really exist – and this was not achieved until a quarter of a century after those preliminary experiments at Green Bank.

There had been earlier indications. In 1983 IRAS, the Infra-Red Astronomical Satellite, detected clouds of cool, possibly planet-forming, material around several stars – notably the brilliant Vega. In one case, that of the southern star, Beta Pictoris, the disk of material was photographed visually from the Las Campanas Observatory in Chile, and there seemed no reason to doubt that planets might exist. Final proof came in 1995, thanks to the patient work of two Swiss astronomers, Michel Mayor and Didier Queloz, using the 1.9-m (76-inch) reflector at the Haute-Provence Observatory in France.

Even with our present-day equipment we cannot manage a direct view of an extra-solar planet; a planet is relatively small, shines only by reflected light and is overpowered by the glare from its parent star. However, a planet of sufficient mass could make the parent star 'wobble' very slightly, and this would show up by Doppler shifts of the lines in the star's spectrum. This is precisely what Mayor and Queloz found with the 5.5-magnitude star 51 Pegasi, which is 42 light-years away and of the same type as the Sun and only marginally less luminous. The real surprise was that the planet appeared to have a mass about half that of Jupiter, and was a mere 7.4 million km (4.6 million miles) from the star, completing one orbit in just over 4 days. With that sort of mass, it had to be a gas-giant, but its nearness to the star was totally unexpected. The surface temperature was estimated at being 1300° C, hotter than many brown dwarfs and much too hot to allow for any form of life.

More discoveries followed, and at the present time the total of known extra-solar planets exceeds 200. Most of them are 'hot jupiters', but there seems every reason to believe that planets similar to the Earth must also exist; it is simply that for obvious reasons they are much more difficult to detect – the 'wobbles' produced in the motion of the parent star are so slight. Another method is to check on the brightness of a star and see whether there are any periodical dips in magnitude, due to an orbiting planet passing between the star and ourselves and blocking out part of the star's light. This 'transit method' has already given good results and is to be extended in the near future with what is called Project Kepler. A large telescope with ultra-sensitive equipment will be sent up in a space-craft and will monitor about 100 000 solar-type stars.

Some stars are now known to have true planetary systems; thus Upsilon Andromedae, 44 light-years away, seems to have three planets, none of which is less massive than Jupiter. Of special interest is Epsilon Eridani, just under 11 light-years from us, and one of the two closest stars of the same type as the Sun (Tau Ceti is the other). The planet of Epsilon Eridani has a mass rather less than that of Jupiter, and moves around the star at a distance of 494 million km (307 million miles) in a period of almost seven years. Clearly it is a gas-giant – but why should there not be Earth-type planets closer in?

If planets are commonplace in the Galaxy – and there is now no doubt about this – it is reasonable to assume that there must be life. To suggest that our Earth is unique seems illogical. There are still a few astronomers who maintain that we are alone in the universe, but the majority view is that there is every chance that at this moment, somewhere in the Galaxy, there is an intelligent being who is considering the same problem and wondering if there can be life on a planet moving around an undistinguished yellow dwarf star – the star that we call the Sun.

Of course, we can do little more than speculate, but there may well be races of all kinds in the Galaxy, from Stone Age cultures through to brilliant civilizations far ahead of ours, both morally and technologically. Yet we have to admit that there may also be ruined, radioactive planets whose inhabitants have wiped each other out. Let us hope that this is a fate which we on Earth will be sensible enough to avoid. Recall the sage words of Percival Lowell, of Martian canal fame, written 100 years ago:

'War is a survival among us from savage times, and affects now the boyish and unthinking element of the nation. The wisest realize that there are better ways for practising heroism and other and more certain ways of ensuring the survival of the fittest. It is something a people outgrow. But whether they consciously practise peace or not, nature in its evolution eventually practises it for them, and after enough inhabitants of a globe have killed each other off, the remainder must find it more advantageous to work together for the common good.'

There is always a chance that some other civilization has mastered the art of interstellar travel and is capable of paying us a visit. There is not the slightest evidence that this has happened yet (despite the flying saucer craze, which began in the 1940s and still lingers on), but it is not out of the question. If it ever happens, there will be no cause for alarm, because any civilization advanced enough to travel between the stars will have long since 'grown up', to use Lowell's term. Frankly, I would welcome such a visit and I am sure that such a civilization could teach us a great deal!

So far as radio communication is concerned, the only possible procedure is to use the language of mathematics. We did not invent mathematics; we merely discovered it, and it is universal. If we receive a signal which is rhythmical enough to be classed as non-natural, we will have proved our point.

The time delay is always going to be a problem. Transmit a message to, say, Epsilon Eridani in 2008, and it will arrive in 2016. Assuming that an obliging radio astronomer there hears it and sends a prompt reply, we will receive our answer in 2028 – a total delay of 20 years. Unfortunately, there seems to be nothing we can do about this.

So far as interstellar travel is concerned, what are the possibilities of exotic forms of travel to which the finite speed of light is no barrier? At the moment we have no idea how this might be achieved. It will require a fundamental breakthrough which may come either in the foreseeable future, the far future, or never. Science fiction does have a habit of turning into science fact, and certainly television or even radio would have seemed hopelessly fantastic only a few centuries ago, but so far as we are concerned, speculation about space-warps, time-warps, thought-travel and similar methods is not only endless but also rather pointless. By 3008, the situation may be very different!

I hope that this book has interested you. I have done no more than try to provide an introduction to astronomy, and much has had to be left out. However, if you feel encouraged enough to take matters further, then I will be well satisfied.

glossary

absolute magnitude The apparent magnitude which a star would have if it could be viewed from a standard distance of 10 parsecs, or 32.6 light-years.

albedo The reflecting power of a planet or other body. For example, the average albedo of the Moon is 7 per cent, so that it reflects 7 per cent of the sunlight falling upon it.

aphelion The furthest distance of a planet or other body from the Sun in its orbit.

apogee The furthest point of the Moon from the Earth in its orbit. The term also applies to any other body orbiting the Earth, such as an artificial satellite.

apparent magnitude The apparent brightness of a celestial body. The lower the magnitude, the brighter the object.

astronomical unit The mean distance of the Earth from the Sun. It is equal to 149 598 500 kilometres (92 976 000 miles).

aurora Aurorae are polar lights; aurora borealis in the northern hemisphere, aurora australis in the southern. They are due to charged particles from the Sun which enter the upper air.

azimuth The bearing of an object in the sky, from north (0 degrees) through east, south and west.

barycentre The centre of gravity of the Earth–Moon system. Since the Earth is 81 times as massive as the Moon, the barycentre lies well inside the Earth's globe.

binary star A stellar system made up of two (or more) components which are physically associated.

black hole A region around a very small, super-massive collapsed star from which not even light can escape.

caldera A very large volcanic crater.

Cassegrain reflector A reflecting telescope in which the secondary mirror is convex, and the light is passed to the eyepiece via a hole in the centre of the main mirror.

celestial sphere An imaginary sphere surrounding the Earth, whose centre is coincident with that of the Earth's globe.

cepheid A short-period, regular variable star.

chromosphere That part of the Sun's atmosphere which lies above the bright surface or photosphere. The dark Fraunhofer lines of the solar spectrum are produced in the chromosphere.

circumpolar star A star which never sets from the latitude of the observer.

coelostat A clock-driven mirror on an axis parallel to the Earth's, so as to reflect the same region of the sky continuously.

colures Great circles on the celestial sphere.

coma The nebulous envelope of the head of a comet.

conjunction 1. The close approach in the sky of two celestial bodies. 2. For the inferior planets, inferior conjunction occurs when the planet passes approximately between the Earth and the Sun, and superior conjunction when it is directly on the far side of the Sun. Planets beyond the Earth's orbit can reach superior conjunction only.

corona The outermost part of the Sun's atmosphere.

cosmic rays High-speed atomic particles reaching the Earth from outer space.

culmination The maximum altitude of a celestial body above the horizon.

day, sidereal The interval between successive culminations of the same star: 23h 56m 4.091s.

day, solar The mean interval between successive culminations of the Sun; it is longer than the sidereal day because of the Sun's eastward motion. Its mean length is 24h 3m 56.555s.

declination The angular distance of a celestial body north or south of the celestial equator.

dichotomy The exact half-phase of the Moon or an inferior planet.

Doppler effect The apparent change in wavelength of a light-source according to its motion relative to the observer.

earthshine The faint luminosity of the non-sunlit part of the Moon, due to light reflected on to the Moon from the Earth.

eclipse, lunar The passage of the full moon through the shadow cast by the Earth. Lunar eclipses may be total, partial or penumbral.

eclipse, solar The occultation of the Sun by the Moon. Solar eclipses may be total, partial or annular (an annular eclipse occurs with the Moon in the far part of its orbit, so that its disk then appears smaller than that of the Sun).

ecliptic The apparent yearly path of the Sun among the stars, i.e. the projection of the Earth's orbit on to the celestial sphere.

equator, celestial The projection of the Earth's equator on to the celestial sphere: declination 0 degrees.

equatorial mounting for a telescope A mounting in which the telescope is set up on a polar axis that is parallel to the axis of rotation of the Earth.

equinox The points where the ecliptic cuts the celestial equator. The vernal equinox (First Point of Aries) is now in Pisces, and the autumnal equinox (First Point of Libra) is in Virgo. The Sun crosses the vernal equinox in late March and the autumnal equinox in late September.

escape velocity The minimum velocity needed for a body to escape from the surface of a planet (or other body) without extra impetus; in the case of the Earth, 11.2 km (7 miles) per second.

extinction The apparent reduction in light of a celestial body depending upon its altitude above the horizon.

eyepiece (or ocular) The lens, or combination of lenses, at the eye-end of a telescope, responsible for all the actual magnification of the image.

faculae Bright patches on the surface of the Sun.

flare, solar A brilliant, energetic outbreak above the Sun's surface.

galaxies Systems of stars.

gamma-rays Ultra-short-wave radiations.

Gegenschein (English, Counterglow) A faint glow in the sky exactly opposite to the Sun. It is due to the illumination of interplanetary material in the main plane of the Solar System.

gibbous phase The phase of the Moon or a planet between half and full.

great circle A circle on the surface of a sphere whose plane passes through the centre of that sphere.

Gregorian reflector An obsolete form of reflecting telescope which has a concave secondary mirror, the light being passed to the eyepiece through a hole in the centre of the main mirror.

Hertzsprung–Russell (HR) diagram A diagram in which stars are plotted according to their spectral types and their luminosities (see Figure 15.1).

hour angle The time which has elapsed since the culmination of a celestial object.

hour circle A great circle on the celestial sphere which passes through both the celestial poles.

inferior planets Mercury and Venus, which are closer to the Sun than we are.

infra-red radiation Light of wavelength longer than that of visible light, but shorter than that of radio waves.

Julian day A count of the days, reckoned from 1 January 4713 BC.

kiloparsec 1000 parsecs (3260 light-years).

libration The apparent 'tilting' of the Moon as seen from the Earth.

light-year The distance travelled by light in one year: 9.4607 million million km (5.880 million million miles).

Local Group The group of galaxies of which our Galaxy is a member.

lunation The interval between successive new moons: 29d 12h 44m.

magnetosphere The region of the magnetic field around a celestial body.

Main Sequence A band across the Hertzsprung–Russell diagram running from top left to lower right. It includes the vast majority of stars.

Maksutov telescope A telescope involving both mirrors and lenses.

megaparsec 1 million parsecs.

meridian, celestial The great circle on the celestial sphere which passes through the zenith and both celestial poles.

meteor A piece of cometary débris; a small particle which burns away in the Earth's upper air, producing the effect known as a shooting-star.

meteorite A solid body which lands on Earth. Meteorites come from the asteroid belt, and are not associated with comets or meteors.

micron (μ) One-thousandth of a millimetre.

minor planet (or asteroid) A small planetary body; most asteroids move in the region of the Solar System between the orbits of Mars and Jupiter.

nebula A cloud of dust and gas in space.

neutron A fundamental particle with no electrical charge, but with a mass equal to that of a proton.

neutron star The remnant of a supernova outburst; the core of a disrupted star, now made up of neutrons.

neutrino A particle with no electrical charge and no mass (or at least very little).

Newtonian reflector A reflecting telescope in which the light is collected by a curved mirror, and directed to the eyepiece via a secondary flat mirror inclined at an angle of 45 degrees.

nodes The points at which the orbit of the Moon, a planet or a comet cuts the plane of the ecliptic.

nova The sudden, temporary outburst of a formerly faint star. All novae are binary systems.

obliquity of the ecliptic The angle between the ecliptic and the celestial equator: 23°26′45″.

occultation The covering-up of one celestial body by another.

opposition The position of a planet when exactly opposite to the Sun in the sky.

orbit The path of a celestial body.

parallax, trigonometrical The apparent shift of an object against its background when viewed from different directions.

parsec The distance at which a star would have a parallax of 1 second of arc: 3.26 light-years. (No star apart from the Sun is as close as this.)

penumbra 1. The area of partial shadow to either side of the main cone of shadow cast by the Earth. 2. The lighter part of a sunspot.

perigee The position of the Moon when closest to the Earth in its orbit. The term also applies to any other body in Earth orbit, such as an artificial satellite.

perihelion The position of a planet or other body when closest to the Sun in its orbit.

photon The smallest 'unit' of light.

photosphere The bright surface of the Sun.

planetary nebula A small, dense, hot star surrounded by a shell of gas. It is neither a planet nor a true nebula!

poles, celestial The north and south points of the celestial sphere.

position angle The apparent direction of one object with reference to another, measured from the north point of the main object through east, south and west.

precession The apparent slow movement of the celestial poles, due to the changing direction of the Earth's axis of rotation.

prime meridian The meridian on the Earth's surface which passes through Greenwich Observatory, England. It marks longitude 0 degrees.

prominences Masses of glowing hydrogen gas rising from the bright surface of the Sun.

proper motion, stellar The individual motion of a star on the celestial sphere.

proton A fundamental particle with unit positive electrical charge.

pulsar A rotating neutron star, emitting pulsed radio waves.

quadrature The position of the Moon, planet or other body when at right angles to the Sun as seen from the Earth.

quasar A very remote, super-luminous object; the nucleus of a very active galaxy.

radial velocity The towards-or-away movement of a celestial body relative to the observer.

radiant The point in the sky from which the meteors of any particular shower appear to diverge.

radius vector A straight line between a fixed point and a variable point.

retardation The difference in the time of moonrise from one night to the next.

retrograde motion Orbital or rotational movement in a sense opposite to that of the Earth's axial or orbital rotation.

right ascension The angular distance of a celestial body from the vernal equinox, measured eastwards. It is usually given in units of time: the interval between the culmination of the vernal equinox, and the culmination of the object.

Schmidt telescope An instrument which collects its light by means of a spherical mirror; a correcting plate is placed at the top of the tube.

Seyfert galaxies Galaxies with relatively small, bright nuclei and weak spiral arms. Most of them are powerful radio emitters.

sidereal period The revolution period of a planet or other body around the Sun, or of a satellite around a planet.

sidereal time The local time reckoned according to the apparent rotation of the celestial sphere. When the vernal equinox crosses the observer's meridian, the sidereal time is 0 hours.

solar wind A flow of particles streaming out constantly from the Sun.

solstices The times when the Sun is at its maximum declination: 23½ degrees north or south.

spectroscopic binary A binary system whose components are too close together to be seen separately from Earth, but which can be tracked down by the Doppler shifts in their spectra.

speculum The main mirror of a reflecting telescope.

superior planet A planet whose orbit is further from the Sun than that of the Earth.

supernova A colossal stellar outburst, involving (I) the destruction of the white dwarf component of a binary system, or (II) the collapse of a very massive star. A supernova of the second type usually produces a neutron star, which emits pulsed radio waves and is termed a pulsar.

synodic period The interval between successive oppositions of a superior planet.

syzygy The position of the Moon in its orbit when new or full.

tektites Small, glassy objects found in restricted areas of the Earth. They are probably of terrestrial origin.

terminator The boundary between the sunlit and night sides of the Moon or a planet.

transit 1. The passage of a celestial body across the observer's meridian. 2. The passage of an inferior planet across the face of the Sun.

umbra 1. The main cone of shadow cast by the Earth. 2. The darkest part of a sunspot.

Van Allen zones Zones of charged particles surrounding the Earth.

variable stars Stars which change in output over relatively short periods.

white dwarf A very small, dense, highly-evolved star.

zenith The observer's overhead point.

zodiac The belt stretching around the sky, 8 degrees to either side of the ecliptic, in which the Sun, Moon and principal planets are always to be found.

zodiacal light A cone of light stretching upward from the horizon and extending along the ecliptic. It is due to sunlight illuminating interplanetary material in the main plane of the Solar System.

Zürich number A count of the sunspot activity. It is given by the formula $Z = k (10g + f)$, where Z is the Zürich number, g is the number of groups seen, f is the number of individual spots, and k is a constant depending on the observer's equipment; it is usually near 1. Zürich numbers are also known as Wolf numbers; the system was devised by R. Wolf of Zürich in 1852.

Appendix 1 Planetary data

Planet	Mean distance from Sun, millions of km (miles)	Orbital period	Axial rotation period	Inclination of axis degrees
Mercury	58 (36)	88 days	58.6 days	2
Venus	108 (67)	224.7 days	243 days	178
Earth	150 (93)	365.2 days	23h 56m	23.4
Mars	228 (141.5)	687 days	24h 37m	24
Jupiter	777 (483)	11.9 years	9h 50m	3
Saturn	1426 (886)	29.5 years	10h 14m	26
Uranus	2869 (1783)	84 years	17h 14m	98
Neptune	4494 (2793)	164.8 years	16h 7m	29

Planet	Orbital eccentricity	Orbital inclination, degrees	Maximum magnitude	No. of satellites
Mercury	0.206	7.0	−1.9	0
Venus	0.007	3.4	−4.4	0
Earth	0.017	0	−	1
Mars	0.093	1.9	−2.8	2
Jupiter	0.048	1.3	−2.6	39
Saturn	0.056	2.5	−0.3	30
Uranus	0.047	0.8	+5.6	20
Neptune	0.009	1.8	+7.7	8

Planet	Diameter, km (miles) (equatorial)	Escape velocity km (miles)/sec	Surface gravity, Earth = 1	Mass, Earth = 1	Mean surface temp, ° C
Mercury	4875 (3030)	4.2 (2.6)	0.38	0.06	+427
Venus	12 105 (7523)	10.3 (6.4)	0.90	0.86	+480
Earth	12 753 (7926)	11.3 (7.0)	1	1	+22
Mars	6794 (4222)	5.1 (3.2)	0.38	0.11	−23
Jupiter	143 883 (89 424)	59.5 (37.00)	2.64	318	−150
Saturn	120 537 (74 914)	35.4 (22.00)	1.16	95	−180
Uranus	51 118 (31 770)	22.5 (14.00)	1.17	15	−214
Neptune	50 539 (31 410)	24.1 (15.00)	1.2	17	−220

Appendix 2
The principal planetary satellites

The principal satellites are listed in the table overleaf. Those with retrograde motion are distinguished by an asterisk*.

Mars Phobos, Deimos

Jupiter Metis, Adrastea, Amalthea, Thebe, Io, Europa, Ganymede, Callisto, Leda, Himalia, Lysithea, Elara, Anake*, Carme*, Pasiphaë*, Sinope*.

Saturn Pan Atlas, Prometheus, Pandora, Epimetheus, Janus, Mimas, Enceladus, Tethys, Telesto, Calypso, Dione, Helene, Rhea, Titan, Hyperion, Iapetus, Phoebe*.

Uranus Cordelia, Ophelia, Bianca, Cressida, Desdemona, Juliet, Portia, Rosalind, Belinda, Puck, Miranda, Ariel, Umbriel, Titania, Oberon, Caliban, Stephano, Sycorax, Prospero, Setebos.

Neptune Naiad, Thalassa, Despina, Galatea, Larissa, Proteus, Triton*, Nereid.

Name	Mean distance from centre of primary, thousands of km (miles)	Orbital period d h m	Diameter km (miles) (max)	Magnitude
EARTH				
Moon	385 (239)	27 7 43	3475 (2160)	−12.7 (full)
MARS				
Phobos	9.3 (5.8)	0 7 39	27 x 23 x 18 (17 x 14 x 11)	11.6
Deimos	23.5 (14.6)	1 6 18	31 x 27 x 26 (19 x 17 x 16)	12.8
JUPITER				
Amalthea	182 (113)	0 11 57	262 x 146 x 143 (163 x 91 x 89)	14.1
Io	422 (262)	1 18 28	3643 (2264)	5.0
Europa	671 (417)	3 13 14	3130 (1945)	5.3
Ganymede	1072 (666)	7 3 43	5268 (3274)	4.6
Callisto	1883 (1170)	16 16 32	4796 (2981)	5.6
SATURN				
Mimas	185 (115)	0 22 37	420 (261)	12.9
Enceladus	238 (148)	1 8 53	512 (318)	11.8
Tethys	294 (183)	1 21 18	1046 (650)	10.3
Dione	378 (235)	2 17 41	1120 (696)	10.4
Rhea	528 (328)	4 12 25	1529 (950)	9.7
Titan	1223 (760)	15 22 41	5150 (3201)	8.3
Hyperion	1480 (920)	21 6 38	360 (224)	14.2
Iapetus	3540 (2200)	79 7 56	1435 (892)	10var
Phoebe	12 952 (8050)	550 10 50	230 (143)	16.5
URANUS				
Miranda	130 (81)	1 19 50	471 (293)	16.5
Ariel	191 (119)	2 12 29	1158 (720)	14.4
Umbriel	267 (166)	4 3 28	1170 (727)	15.3
Titania	438 (272)	8 16 56	1578 (981)	14.0
Oberon	587 (365)	13 11 7	1524 (947)	14.2
NEPTUNE				
Triton	354 (220)	5 21 3	2705 (1681)	13.6
Nereid	5857 (3640)	359 21 7	240 (149)	18.7

Trans-Neptunian dwarf planets

	Mean distance from Sun, millions of km (miles)	Orbital Period, y	Eccentricity	Inclination, degrees	Rotation period	Diameter, km (miles)	
134340 Pluto	5899 (3666)	248	0.248	17.1	6d 9 h	2306 (1433)	Kuiper Belt
136199 Eris	10 121 (6290)	557	0.441	44.2	?	2414 (1500)	Scattered Disk Object

(Main Belt asteroid 1 Ceres is also ranked as a dwarf planet)

Appendix 3 Selected asteroids

	Mean distance from Sun, millions of km (miles)	Orbital period, years	Diameter, km (miles)	Rotation period, hours	Magni- tude
1 Ceres	413.5 (257.0)	4.61	940 (584)	9.1	7.4
2 Pallas	414.2 (257.4)	4.62	579 x 470 (360 x 292)	7.8	8.0
3 Juno	398.7 (247.8)	4.36	288 x 230 (179 x 143)	7.2	8.7
4 Vesta	352.9 (219.3)	3.63	576 (358)	5.3	6.5
5 Astraea	385.0 (239.3)	4.14	121 (75)	16.8	9.8
10 Hygeia	470.8 (292.6)	5.59	430 (267)	17.5	10.2
243 Ida	418.9 (260.4)	4.84	39 x 34 (24 x 21)	4.6	14.6
279 Thule	635.2 (394.8)	8.23	130 (81)	7.4	15.4
433 Eros	268.5 (166.9)	1.76	29 (18)	5.3	8.6 (max)
588 Achilles	756 (470)	11.77	116 (72)	58	15.3
944 Hidalgo	587 (365)	14.15	27 (17)	10	15.3
951 Gaspra	301 (187)	3.28	24 (15)	20	14.1
1221 Amor	225 (140)	2.66	1.6 (1)		18
1566 Icarus	193 (120)	1.12	1.6 (1)	2.3	18
1685 Toro	159 (99)	1.60	8 (5)	7.6	10.2
1862 Apollo	159 (99)	1.78	1.6 (1)	1.4	17
2062 Aten	130 (81)	0.95	1.6 (1)		18
2100 Ra-Shalom	98 (61)	0.76	0.5 (0.3)	20	19
2340 Hathor	98 (61)	0.77	0.5 (0.3)		19
3200 Phaethon	106 (66)	1.43	5 (3)	4	17

Appendix 4 Selected comets

Name	Period, years	Distance from Sun, millions of miles:	
		min	max
Encke	3.3	52 (32)	611 (380)
Grigg–Skjellerup	5.1	150 (93)	740 (460)
D'Arrest	6.2	177 (110)	837 (520)
Pons–Winnecke	6.3	187 (116)	837 (520)
Giacobini–Zinner	6.5	148 (92)	885 (550)
Finlay	6.9	164 (102)	933 (580)
Faye	7.4	241 (150)	901 (560)
Tuttle	13.3	153 (95)	1561 (970)
Crommelin	27.9	111 (69)	2639 (1640)
Tempel–Tuttle	32.9	146 (91)	2928 (1820)
Halley	76.1	88 (55)	5294 (3290)
Swift–Tuttle	135	143 (89)	7723 (4800)

Principal meteor showers

Name	Begins	Max	Ends	ZHR*
Quadrantids	1 Jan	4 Jan	6 Jan	60
Lyrids	19 Apr	21 Apr	25 Apr	10
Perseids	23 July	12 Aug	20 Aug	75
Orionids	16 Oct	22 Oct	27 Oct	25
Taurids	20 Oct	3 Nov	30 Nov	10
Leonids	15 Nov	17 Nov	20 Nov	var.
Geminids	7 Dec	13 Dec	16 Dec	75

*ZHR: Zenithal Hourly Rate

Appendix 5 The constellations

Large and important constellations are given in capital letters. Hemisphere indicates the position of the main part of the constellation. Z indicates Zodiacal.

Name	English name	1st-magnitude star(s)	Hemisphere
ANDROMEDA	Andromeda	–	N
Antlia	The Air-pump	–	S
Apus	The Bird of Paradise	–	S
AQUARIUS	The Water-bearer	–	S Z
AQUILA	The Eagle	Altair	N
Ara	The Altar	–	S
ARIES	The Ram	–	N Z
AURIGA	The Charioteer	Capella	N
BOÖTES	The Herdsman	Arcturus	N
Caelum	The Graving Tool	–	S
Camelopardalis	The Giraffe	–	N
CANCER	The Crab	–	N Z
Canes Venatici	The Hunting Dogs	–	N
CANIS MAJOR	The Great Dog	Sirius	S
Canis Minor	The Little Dog	Procyon	N
CAPRICORNUS	The Sea-Goat	–	S Z
CARINA	The Keel	Canopus	S
CASSIOPEIA	Cassiopeia	–	N
CENTAURUS	The Centaur	Alpha Centauri, Agena	S
Cepheus	Cepheus	–	N
CETUS	The Whale	–	S
Chamaeleon	The Chameleon	–	S
Circinus	The Compasses	–	S
Columba	The Dove	–	S
Coma Berenices	Berenice's Hair	–	N
Corona Australis	The Southern Crown	–	S
Corona Borealis	The Northern Crown	–	N
Corvus	The Crow	–	S
Crater	The Cup	–	S
CRUX AUSTRALIS	The Southern Cross	Acrux, Beta Crucis	S
CYGNUS	The Swan	Deneb	N
Delphinus	The Dolphin	–	N
Dorado	The Swordfish	–	S

DRACO	The Dragon	–	N
Equuleus	The Foal	–	N
ERIDANUS	The River	Achernar	S
Fornax	The Furnace	–	S
GEMINI	The Twins	Pollux	N
Grus	The Crane	–	S
HERCULES	Hercules	–	N
Horologium	The Clock	–	S
HYDRA	The Watersnake	–	S
Hydrus	The Little Snake	–	S
Indus	The Indian	–	S
Lacerta	The Lizard	–	N
LEO	The Lion	Regulus	N
Leo Minor	The Little Lion	–	N
Lepus	The Hare	–	S
LIBRA	The Balance	–	S Z
Lupus	The Wolf	–	S
Lynx	The Lynx	–	N
LYRA	The Lyre	Vega	N
Mensa	The Table	–	S
Microscopium	The Microscope	–	S
Monoceros	The Unicorn	–	Equatorial
Musca Australis	The Southern Fly	–	S
Norma	The Rule	–	S
Octans	The Octant	–	S
OPHIUCHUS	The Serpent-bearer	–	Equatorial
ORION	Orion	Rigel, Betelgeux	Equatorial
Pavo	The Peacock	–	S
PEGASUS	The Flying Horse	–	N
PERSEUS	Perseus	–	N
Phoenix	The Phoenix	–	S
Pictor	The Painter	–	S
Pisces	The Fishes	–	N Z
Piscis Australis	The Southern Fish	Fomalhaut	S
PUPPIS	The Poop	–	S
Pyxis	The Mariner's Compass	–	S
Reticulum	The Net	–	S
Sagitta	The Arrow	–	N
SAGITTARIUS	The Archer	–	S Z
SCORPIUS	The Scorpion	Antares	S Z
Sculptor	The Sculptor	–	S
Scutum	The Shield	–	S
Serpens	The Serpent	–	N
Sextans	The Sextant	–	S
TAURUS	The Bull	Aldebaran	N

Telescopium	The Telescope	–	S
Triangulum	The Triangle	–	N
Triangulum Australe	The Southern Triangle	–	S
Tucana	The Toucan	–	S
URSA MAJOR	The Great Bear	–	N
URSA MINOR	The Little Bear	–	N
VELA	The Sails	–	S
VIRGO	The Virgin	Spica	Equatorial
Volans	The Flying Fish	–	S
Vulpecula	The Fox	–	N

Appendix 6 Stars of the first magnitude

		Spectrum	Magnitude	Declination°
Alpha Canis Majoris	Sirius	A	–1.5	–17
Alpha Carinae	Canopus	F	–0.7	–53
Alpha Centauri		K+G	–0.3	–61
Alpha Boötis	Arcturus	K	–0.0	+19
Alpha Lyrae	Vega	A	0.0	+39
Alpha Aurigae	Capella	G	0.1	+46
Beta Orionis	Rigel	B	0.1	–08
Alpha Canis Minoris	Procyon	F	0.4	+02
Alpha Eridani	Achernar	B	0.5	–57
Alpha Orionis	Betelgeux	M	var.	+07
Beta Centauri	Agena	B	0.6	–60
Alpha Aquilae	Altair	A	0.8	+09
Alpha Crucis	Acrux	B+B	0.8	–63
Alpha Tauri	Aldebaran	K	0.8	+17
Alpha Scorpii	Antares	M	1.0	–26
Alpha Virginis	Spica	B	1.0	–11
Beta Geminorum	Pollux	K	1.1	+28
Alpha Piscis Australis	Fomalhaut	A	1.2	–30
Alpha Cygni	Deneb	A	1.2	+45
Beta Crucis		B	1.2	–60
Alpha Leonis	Regulus	B	1.3	+12

Next in order come Epsilon Canis Majoris (1.5), and Alpha Geminorum (Castor), Gamma Crucis, Lambda Scorpii, Gamma Orionis and Beta Tauri (all 1.6).

Appendix 7 Proper names of stars

Very few stars below the first magnitude have proper names which are in general use, but the following are found often enough to be worth listing here.

Alpha Andromedae	Alpheratz
Epsilon Boötis	Izar
Alpha Canum Venaticorum	Cor Caroli
Epsilon Canis Majoris	Adhara
Alpha Cassiopeiae	Shedir
Beta Ceti	Diphda
Omicron Ceti	Mira
Alpha Coronae Borealis	Alphekka
Beta Cygni	Albireo
Alpha Draconis	Thuban
Theta Eridani	Acamar
Alpha Geminorum	Castor
Gamma Geminorum	Alhena
Alpha Herculis	Rasalgethi
Alpha Hydrae	Alphard
Beta Leonis	Denebola
Gamma Leonis	Algieba
Alpha Ophiuchi	Rasalhague
Gamma Orionis	Bellatrix
Delta Orionis	Mintaka
Epsilon Orionis	Alnilam
Zeta Orionis	Alnitak
Kappa Orionis	Saiph
Beta Pegasi	Scheat
Alpha Persei	Mirphak
Beta Persei	Algol
Lambda Scorpii	Shaula
Upsilon Scorpii	Lesath
Beta Tauri	Alnath
Eta Tauri	Alcyone
Alpha Ursae Majoris	Dubhe
Beta Ursae Majoris	Merak
Gamma Ursae Majoris	Phad
Delta Ursae Majoris	Megrez
Epsilon Ursae Majoris	Alioth
Zeta Ursae Majoris	Mizar
Eta Ursae Majoris	Alkaid
80 Ursae Majoris	Alcor
Alpha Ursae Minoris	Polaris
Beta Ursae Minoris	Kocab

taking it further

Further reading

Arnold, H., Doherty, P. and Moore, P. *Photographic Atlas of the Stars* (Institute of Physics Publishing, 2000).

Levy, D. *David Levy's Guide to Observing Meteor Showers* (Cambridge University Press, 2003).

Maunder, M. *Lights in the Sky* (Springer, 2007).

May, Brian, Moore, P. and Lintot, C. *Bang! The Complete History of the Universe* (Carlton Books, 2008).

Moore, P. *Atlas of the Universe* (George Philip, 2000).

Moore, P. *Venus* (Cassell, 2006).

Nicolson, I. *Unfolding our Universe* (Cambridge University Press, 1999).

Nicolson, I. *The Dark Side of the Universe* (Canopus, 2007).

Parker, G. *Making Beautiful Images of the Sky* (Springer, 2007).

Ratledge, D. *Observing the Caldwell Objects* (Springer, 2000).

Tritton, K. *Earth, Life and the Universe* (Curved Air Publications, 2002).

Various authors: *Yearbook of Astronomy* (Macmillan and Co, annually in October). This includes a list of British astronomical societies.

Assorted astronomy and space science websites

NASA homepage
http://www.nasa.gov

Space Telescope Science Institute
http://oposite.stsci.edu/pubinfo

Astronomy Picture of the Day
http://antwrp.gsfc.nasa.gov/apod/astropix.html

European Southern Observatory
http://www.eso.org

National Optical Astronomy Observatories
http://www.noao.edu/outreach

National Space Science Data Center
http://nssdc.gsfc.nasa.gov

SETI Institute
http://www.seti.org/

Jodrell Bank
http://www.jb.man.ac.uk

Jet Propulsion Laboratory Photojournal
http://photojournal.jpl.nasa.gov

Solar and Heliospheric Observatory (SOHO)
http://sohowww.nascom.nasa.gov

Eclipse Information
http://sunearth.gsfc.nasa.gov

Isaac Newton Group of Telescopes
http://www.ing.iac.es

Institute of Astronomy, Cambridge
http://www.ast.cam.ac.uk

Anglo-Australian Observatory
http://www.aao.gov.au/

Sky and Telescope
http://www.skyandtelescope.org

Bridge
British Citizenship Test, The
British Empire, The
British Monarchy from Henry
  VIII, The
Buddhism
Bulgarian
Bulgarian Conversation
Business French
Business Plans
Business Spanish
Business Studies
C++
Calculus
Calligraphy
Cantonese
Caravanning
Car Buying and Maintenance
Card Games
Catalan
Chess
Chi Kung
Chinese Medicine
Christianity
Classical Music
Coaching
Cold War, The
Collecting
Computing for the Over 50s
Consulting
Copywriting
Correct English
Counselling
Creative Writing
Cricket
Croatian
Crystal Healing
CVs
Czech
Danish
Decluttering
Desktop Publishing
Detox
Digital Home Movie Making
Digital Photography
Dog Training
Drawing

Dream Interpretation
Dutch
Dutch Conversation
Dutch Dictionary
Dutch Grammar
Eastern Philosophy
Electronics
English as a Foreign Language
English Grammar
English Grammar as a Foreign
  Language
Entrepreneurship
Estonian
Ethics
Excel 2003
Feng Shui
Film Making
Film Studies
Finance for Non-Financial
  Managers
Finnish
First World War, The
Fitness
Flash 8
Flash MX
Flexible Working
Flirting
Flower Arranging
Franchising
French
French Conversation
French Dictionary
French for Homebuyers
French Grammar
French Phrasebook
French Starter Kit
French Verbs
French Vocabulary
Freud
Gaelic
Gaelic Conversation
Gaelic Dictionary
Gardening
Genetics
Geology
German
German Conversation

German Grammar
German Phrasebook
German Starter Kit
German Vocabulary
Globalization
Go
Golf
Good Study Skills
Great Sex
Green Parenting
Greek
Greek Conversation
Greek Phrasebook
Growing Your Business
Guitar
Gulf Arabic
Hand Reflexology
Hausa
Herbal Medicine
Hieroglyphics
Hindi
Hindi Conversation
Hinduism
History of Ireland, The
Home PC Maintenance and
  Networking
How to DJ
How to Run a Marathon
How to Win at Casino Games
How to Win at Horse Racing
How to Win at Online Gambling
How to Win at Poker
How to Write a Blockbuster
Human Anatomy & Physiology
Hungarian
Icelandic
Improve Your French
Improve Your German
Improve Your Italian
Improve Your Spanish
Improving Your Employability
Indian Head Massage
Indonesian
Instant French
Instant German
Instant Greek

Instant Italian
Instant Japanese
Instant Portuguese
Instant Russian
Instant Spanish
Internet, The
Irish
Irish Conversation
Irish Grammar
Islam
Israeli-Palestinian Conflict, The
Italian
Italian Conversation
Italian for Homebuyers
Italian Grammar
Italian Phrasebook
Italian Starter Kit
Italian Verbs
Italian Vocabulary
Japanese
Japanese Conversation
Java
JavaScript
Jazz
Jewellery Making
Judaism
Jung
Kama Sutra, The
Keeping Aquarium Fish
Keeping Pigs
Keeping Poultry
Keeping a Rabbit
Knitting
Korean
Latin
Latin American Spanish
Latin Dictionary
Latin Grammar
Letter Writing Skills
Life at 50: For Men
Life at 50: For Women
Life Coaching
Linguistics
LINUX
Lithuanian
Magic

Mahjong
Malay
Managing Stress
Managing Your Own Career
Mandarin Chinese
Mandarin Chinese Conversation
Marketing
Marx
Massage
Mathematics
Meditation
Middle East Since 1945, The
Modern China
Modern Hebrew
Modern Persian
Mosaics
Music Theory
Mussolini's Italy
Nazi Germany
Negotiating
Nepali
New Testament Greek
NLP
Norwegian
Norwegian Conversation
Old English
One-Day French
One-Day French – the DVD
One-Day German
One-Day Greek
One-Day Italian
One-Day Polish
One-Day Portuguese
One-Day Spanish
One-Day Spanish – the DVD
One-Day Turkish
Origami
Owning a Cat
Owning a Horse
Panjabi
PC Networking for Small
  Businesses
Personal Safety and Self
  Defence
Philosophy
Philosophy of Mind

Philosophy of Religion
Phone French
Phone German
Phone Italian
Phone Japanese
Phone Mandarin Chinese
Phone Spanish
Photography
Photoshop
PHP with MySQL
Physics
Piano
Pilates
Planning Your Wedding
Polish
Polish Conversation
Politics
Portuguese
Portuguese Conversation
Portuguese for Homebuyers
Portuguese Grammar
Portuguese Phrasebook
Postmodernism
Pottery
PowerPoint 2003
PR
Project Management
Psychology
Quick Fix French Grammar
Quick Fix German Grammar
Quick Fix Italian Grammar
Quick Fix Spanish Grammar
Quick Fix: Access 2002
Quick Fix: Excel 2000
Quick Fix: Excel 2002
Quick Fix: HTML
Quick Fix: Windows XP
Quick Fix: Word
Quilting
Recruitment
Reflexology
Reiki
Relaxation
Retaining Staff
Romanian
Running Your Own Business

Russian
Russian Conversation
Russian Grammar
Sage Line 50
Sanskrit
Screenwriting
Second World War, The
Serbian
Setting Up a Small Business
Shorthand Pitman 2000
Sikhism
Singing
Slovene
Small Business Accounting
Small Business Health Check
Songwriting
Spanish
Spanish Conversation
Spanish Dictionary
Spanish for Homebuyers
Spanish Grammar
Spanish Phrasebook
Spanish Starter Kit
Spanish Verbs
Spanish Vocabulary
Speaking On Special Occasions
Speed Reading
Stalin's Russia
Stand Up Comedy
Statistics
Stop Smoking
Sudoku
Swahili
Swahili Dictionary
Swedish
Swedish Conversation
Tagalog
Tai Chi
Tantric Sex
Tap Dancing
Teaching English as a Foreign
    Language
Teams & Team Working
Thai
Thai Conversation
Theatre

Time Management
Tracing Your Family History
Training
Travel Writing
Trigonometry
Turkish
Turkish Conversation
Twentieth Century USA
Typing
Ukrainian
Understanding Tax for Small
    Businesses
Understanding Terrorism
Urdu
Vietnamese
Visual Basic
Volcanoes, Earthquakes and
    Tsunamis
Watercolour Painting
Weight Control through Diet &
    Exercise
Welsh
Welsh Conversation
Welsh Dictionary
Welsh Grammar
Wills & Probate
Windows XP
Wine Tasting
Winning at Job Interviews
Word 2003
World Faiths
Writing Crime Fiction
Writing for Children
Writing for Magazines
Writing a Novel
Writing a Play
Writing Poetry
Xhosa
Yiddish
Yoga
Your Wedding
Zen
Zulu

geology
david rothery

- Would you like to learn about key geological processes?
- Are you looking for clear explanations?
- Do you want to understand the geological background to climate change?

Geology is a fascinating guide to the nature and history of the Earth. Whether you want to find out why volcanoes and earthquakes happen, where fossils, rocks and minerals come from, or how the Earth has evolved into its present form, this book uncovers all. Extensively illustrated and highly praised, this book will be sure to change the way in which you view the planet you live on.

David Rothery is a volcanologist, geologist, planetary scientist and senior lecturer at the Open University.

teach
yourself

physics
jim breithaupt

- Do you want to understand the key concepts of physics?
- Do you need to know basic physics for a course or exam?
- Do you need to refresh your memory?

Physics offers a comprehensive introduction to the main
branches of physics and the key ideas that run through the
subject. It introduces you to the important concepts and
essential facts, outlines important recent and historic
discoveries, gradually introduces the necessary mathematical
skills and offers questions, answers and worked examples.

Jim Breithaupt is a physics author and lecturer with extensive
teaching experience in schools and colleges.

teach yourself

mathematics
trevor johnson and hugh neill

- Do you need to brush up your maths?
- Do you want to learn key mathematical techniques?
- Are you looking for a clear, accessible guide?

Mathematics will prove invaluable for anyone looking for a step-by-step introduction to key mathematical concepts and techniques. Packed full of worked examples, clear explanations and exercises, this book will guide you through the essentials quickly and painlessly. Covering everything from algebra and geometry to fractions and decimals, you will soon gain both knowledge and confidence.

Trevor Johnson is the chief examiner for Edexcel's International GCSE. **Hugh Neill** is a former A-level examiner. Both are respected authors and consultants.

| teach yourself | **evolution** |
| | james napier |

- Do you want a jargon-free guide to this complex topic?
- Would you like to understand the main arguments?
- Do you want to consider the evidence?

Evolution is a comprehensive guide to this fascinating subject. Outlining the major arguments, evidence and theorists, from Darwin to Dawkins, this book will give you a thorough understanding of a theory that continues to stimulate fierce debate. You will also explore alternative views such as creationism and intelligent design. Putting evolution into a modern context and examining all perspectives, this book will be sure to challenge and engage you.

James Napier teaches biology for the Open University's Science Foundation Course.